Thomas Schulenberg (Ed.)

Annual Report 2017–2018 of the Institute for Nuclear and Energy Technologies

Karlsruhe Institute of Technology

KIT SCIENTIFIC REPORTS 7756

Annual Report 2017 – 2018 of the Institute for Nuclear and Energy Technologies

Edited by
Thomas Schulenberg

KIT Scientific Publishing

Report-Nr. KIT-SR 7756

Impressum

KIT Scientific Publishing

Karlsruher Institut für Technologie (KIT)
KIT Scientific Publishing
Straße am Forum 2
D-76131 Karlsruhe

KIT Scientific Publishing is a registered trademark
of Karlsruhe Institute of Technology.
Reprint using the book cover is not allowed.

www.ksp.kit.edu

Print on Demand 2019 – Gedruckt auf FSC-zertifiziertem Papier

ISSN 1869-9669
ISBN 978-3-7315-0942-4
DOI 10.5445/KSP/1000095726

Content

Institute for Nuclear and Energy Technologies

Structure and Laboratories of the Institute for Nuclear and Energy Technologies

Thomas Schulenberg

The Institute for Nuclear and Energy Technologies (IKET) at KIT addresses selected problems of energy conversion from thermal power to electric power for future power plants, which shall not emit CO_2. Traditionally, these were nuclear power plants, for which IKET concentrates primarily on their safety features and on methods to mitigate severe accidents. Motivated by the nuclear accident in Fukushima, a focus of the research topics within recent years was on the retention of core melt inside the reactor vessel and inside the containment to prevent other power plants from running into similar catastrophes. In its working group on Severe Accident Research, IKET is operating a test facility LIVE for core melt retention inside the vessel, using molten salts to simulate a corium pool in the lower plenum. Fig. 1 shows a view into the open test facility with an inner diameter of 1 m, representing the lower plenum of a reactor. Hot, molten salt can be poured into this vessel, and concentric electric heaters simulate the residual heat of the core melt. The facility can be cooled from outside, such that the melt solidifies to a protecting crust at the vessel walls.

Fig. 1: View into the open test facility LIVE for in-vessel melt retention experiments.

The MOCKA facility, shown Fig. 2, is simulating the interaction of molten corium with concrete inside the containment, using molten zirconium and thermite at realistic temperature and viscosity. The working group on Severe Accident Research is performing these spectacular tests outside, avoiding the risk of hydrogen detonations arising from the decomposition of concrete. Here, the residual heat is added as chemical heat by dropping additional thermite into the test crucible, using the supply system as indicated in Fig. 2. Tests have been performed with different kinds of concrete, with and without rebar, giving the design of future core catchers a solid technical basis.

Fig. 2: MOCKA test of molten corium interaction with concrete

For off-site emergency management, the working group on Accident Management Systems is developing and deploying the decision support system JRODOS, which became the most accepted information system for off-site nuclear emergency management today. Fig. 3 shows a screen shot of the information, which may be given to civil protection organizations in case of a release of radioactive material from a nuclear facility. Colors indicate regions, where an increased gamma dose rate will have to be expected within the next hours, such that evacuation plans or other risk mitigation action can be decided. JRODOS is installed today in 20 European countries and in Asia, expanding its applications every year.

The methodology has been applied meanwhile also to other hazards with severe potential consequences on public health or on the resilience of critical infrastructures in case of a breakdown of power supply.

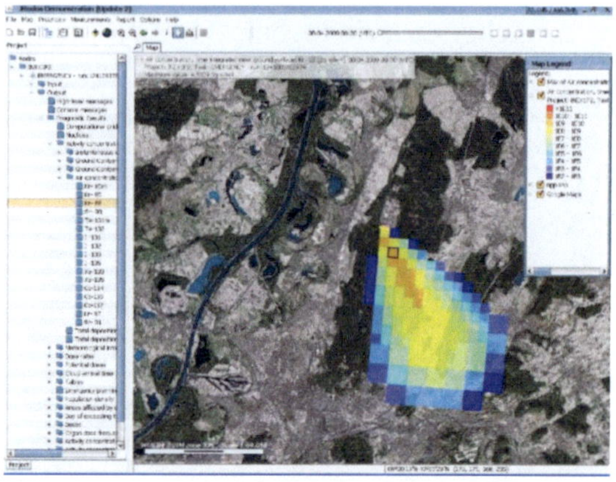

Fig. 3: Total effective gamma dose rate, predicted after a fictive nuclear accident (JRODOS screen shot)

THEADES test facility of the Liquid Metal Laboratory KALLA at IKET, shown in Fig. 4. Up to 40 tons of liquid PbBi can be operated there in a large loop at temperatures up to 400°C to test realistically the mock-up of future nuclear reactor components. The PbBi loop CORRIDA, situated next to the THEADES loop, is operated at elevated temperatures up to 550°C, testing material samples for the Institute for Applied Materials at KIT. In addition, a loop with liquid sodium and one with liquid lead are ready for operation in the same laboratory, providing a comprehensive research infrastructure for all kinds of liquid metal experiments.

Fig. 4: Look into the Liquid Metal Laboratory KALLA with its THEADES test loop.

Besides water-cooled reactors, IKET has been looking as well at innovative nuclear energy systems, either designed for a closed nuclear fuel cycle or for transmutation of spent fuel, both indicating potential alternatives to direct nuclear waste disposal. In this context, generic tests of reactor components for the lead-bismuth cooled research facility MYRRHA, to be built in Belgium, are performed in the

In its Transmutation group, IKET is studying several severe accident scenarios of liquid metal cooled reactors, which are simulated numerically with the SIMMER code in international collaboration. The Institute could thus keep a wider range of nuclear expertise, including advanced neutron physics and reactor kinetics, which is still well recognized in Europe and Japan. For illustration, Fig. 5 shows a predicted fuel particle distribution after a postulated fuel pin failure in the MYRRHA test facility. As the coolant PbBi has a similar density as UO_2, fuel particle may float in the coolant and may be taken away from the reactor core to coolers and pumps. Such predictions are needed during the design phase to protect the facility from severe secondary damage.

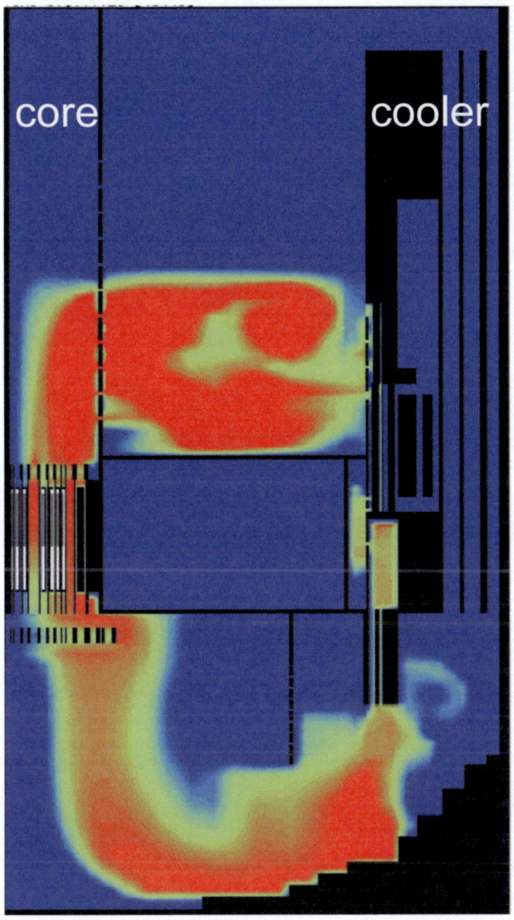

Fig. 5: Fuel particle distribution 90s after a fuel pin failure in MYRRHA

Driven by the German government decision to phase out of nuclear power by 2022, the Institute is working today also on questions of nuclear fusion and on conversion of renewable energies. Its long term experience with liquid metals and with magneto-hydrodynamics enabled studies of flow of PbLi in breeding blankets, which will be exposed to strong magnetic fields in a fusion reactor. The working group on Magneto-Hydrodynamics is operating for this purpose the unique test facility MEKKA with liquid NaK, simulating liquid PbLi, but at room temperature, including its interaction with magnetic fields. Fig. 5 shows a look into the MEKKA laboratory, where a test section connected with the NaK loop is just being inserted into a magnet with up to 3.6 Tesla for magnetohydrodynamic measurements.

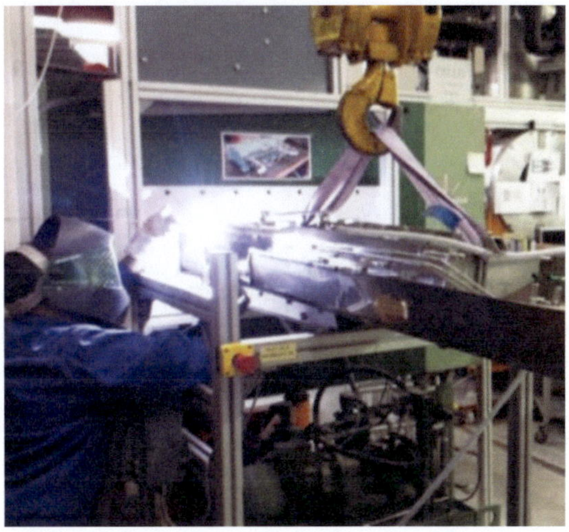

Fig. 6: Insertion of a NaK test loop into the magnet of the MEKKA laboratory.

Liquid PbBi, on the other hand, became a promising coolant for concentrated solar power plants. With its test facility SOMMER, shown in Fig. 7, the Liquid Metal Laboratory is studying the highly effective heat transfer from the focus of a parabolic solar mirror to an energy conversion system using liquid metals as an energy carrier. A heliostat of 30 m^2, visible in the foreground of Fig. 7, is following the sun to illuminate a parabolic mirror of 4m x4 m, visible in the background, which in turn focusses the solar power to a liquid metal receiver inside the building. The focal point could easily melt a steel plate with its peak heat flux of 2.5 MW/m^2, if uncooled, but the PbBi cooled tubes of the receiver are keeping the surface sufficiently cold.

Fig. 7: Test facility SOMMER for concentrated solar power in the Liquid Metal Laboratory.

A high potential for renewable energies, even though economically still rather questionable in Germany, is the use of geothermal energies. For a reasonable energy conversion to electric power, these must be at a temperature of at least 130°C, which is available in Germany only at a depth of several thousand meters. The working group on Energy and Process Engineering at IKET is studying, therefore, such concepts also in international collaboration, e.g. with Indonesia, where even hotter water is found at a significantly lower depth. For effective energy conversion of low temperature energy sources, IKET has developed and built the test facility MONIKA, using propane as a working fluid of a supercritical Rankine cycle. Instead of a geothermal borehole, a thermal power of 1 MW is produced there simply with a boiler to test the facility.

Another interesting application of this process is the conversion of low temperature waste heat to electric power, e.g. from cogeneration power plants in summer times or from chemical facilities. Like with geothermal energies, the challenge is here again to develop a low cost, competitive system. The group is studying also the water chemistry and the physical properties of the geothermal brine with its high concentrations of salt. They predict the occurrence of scale and of released gases in heat exchangers and they developed an in-situ measurement system for continuous monitoring of physical properties.

Fig. 8: Test facility MONIKA with a Rankine cycle of supercritical propane.

A promising carbon free energy carrier for future energy systems is hydrogen. IKET has gained long-term experience with safety aspects of this fuel already since the 1990ies. At these times, hydrogen was rather considered as a safety issue of nuclear power plants during severe accidents, demonstrated impressively by the Fukushima power plant, which exploded due to the production and ignition of hydrogen. Today, hydrogen is often considered and tested as an innovative fuel for automobiles, busses or trains, and the experience of the Hydrogen working group with potential hazards of hydrogen and their mitigation became the basis for safety research for hydrogen driven vehicles. The Hydrogen team has thus supported the installation and operation of a hydrogen fueled shuttle bus at KIT, Fig. 9, and performed tests of innovative hydrogen tanks in its hydrogen test center HYKA. Their code systems GASFLOW and COMB3D are used today worldwide to predict concentrations of released hydrogen in closed buildings like reactor containments, tunnels or garages, and to estimate the mechanical loads on the building structure in case of a detonation.

Another interesting task of this group is the prediction of dust explosions, with and without the additional effects of exploding gases.

Fig. 9: Hydrogen gas station and shuttle bus, supervised by IKET and ProScience.

Can hydrogen be produced from natural gas without CO_2-emissions? The KALLA team has developed and demonstrated indeed an innovative gas reformation process, which decomposes methane into hydrogen and elementary carbon at high temperatures. The chemical reaction is performed in liquid tin, which serves as a catalyst for methane reduction, and solid carbon could be produced at the tin surface, Fig. 10.

Fig. 10: Carbon produced from pyrolysis of methane in the KALLA laboratory.

Prediction and measurement of two-phase flow of steam and liquid has traditionally been a focus of IKET, applicable to all kinds of power plant concepts using the Rankine cycle for energy conversion. In collaboration with a consortium of research and industry partners in the NUBEKS project, the working group on Multiphase Flows has recently performed critical heat flux tests in its COSMOS-L facility at pressures up to 3 bar. The detailed data were taken by the other partners then to validate innovative prediction methods for the occurrence of a boiling crisis. Other numerical modelling activities of IKET are including stratified two-phase flows, steam injectors with direct contact condensation, and heat transfer phenomena close to the critical pressure, which were studied in doctoral theses.

Fig. 11: Evaporator of COSMOS-H, designed for critical heat flux experiments at high pressure.

A large two-phase flow loop COSMOS-H with pressures up to 170 bar has been installed next to the low pressure facility, Fig. 11, being commissioned in near future. The facility has a total heating power of 1.2 MW and can produce a steam/water mass flow up to 1.4 kg/s. An optical access to the test section shall enable laser-optical measurements at high pressure, which might give a completely new insight into the physics of nucleate boiling.

The combination of science and technology with education and training is a systematic approach at KIT, and IKET is contributing accordingly to courses in mechanical engineering, supervises several bachelor and master theses each year and coordinates master programs in energy technologies. Compact courses on energy technologies are given also in executive master programs and in the Framatome Professional School, which is funded by industry and managed by IKET.

An overview of the structure of IKET is given by the organization chart, Fig. 12. Each working group is acting independently in its research field, but they are all supported by a joint infrastructure, comprising a metal workshop, manufacturing urgent test components, a welding shop, and an electromechanical workshop. A view into the metal workshop, Fig. 13, shows that the Institute is well equipped with machines and tools to support the test laboratories most effectively. Indeed, the high number of large test facilities could hardly be operated without this joint collaboration.

Other tasks of the infrastructure include the IT-administration, business administration and public websites. The Infrastructure team is active as well in education and training activities. Every year, at least six students of the Baden-Württemberg Cooperative State University are employed by IKET, managed by the Infrastructure group, to work with the research teams as part of their educational program.

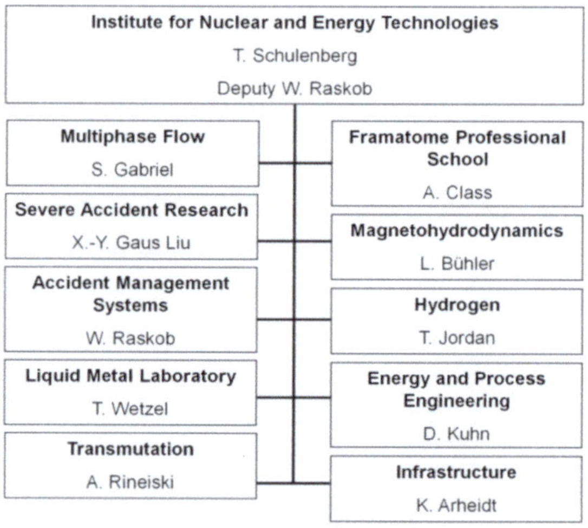

Fig. 12: Organization chart of the Institute for Nuclear and Energy Technologies

Fig. 13: View into the metal workshop of IKET.

By the end of 2018, around 90 scientists, engineers and technicians have been working at IKET on this wide range of CO_2-free technologies for energy conversion. They contribute to the HGF-programs on Nuclear Safety

(NUSAFE), Nuclear Fusion (FUSION), Renewable Energies (EE) as well as Storage and Cross-linked Infrastructures (SCI), with a focus on NUSAFE. Fig. 14 illustrates that more than 2/3 of the IKET employees were still needed in 2018 for tasks in NUSAFE, despite the German phase-out plans of nuclear power.

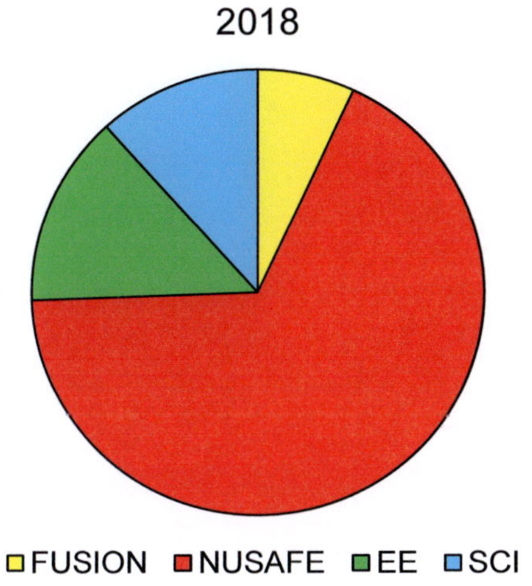

Fig. 14: Assignment of IKET personnel to HGF programs.

Results of the scientific program of IKET have been documented in more than 100 publications per year. Some highlights of recent achievements will be outlined in the following paragraphs. A list of all relevant publications of the Institute can be found in the attachment.

Group: Magnetohydrodynamics

Magnetohydrodynamics for liquid-metal blankets

L. Bühler, H.-J. Brinkmann, V. Klüber, C. Köhly, C. Mistrangelo

Numerical simulations of 3D magnetohydrodynamic flows for dual-coolant lead lithium blankets

The dual coolant lead lithium (DCLL) blanket is one of the liquid metal blanket concepts foreseen in a DEMOnstration nuclear fusion reactor. The liquid metal serves here as breeder, neutron multiplier, neutron shield and as coolant to remove the volumetrically generated heat. The movement of the electrically conducting fluid in the strong magnetic field induces electric currents responsible for strong Lorentz forces, high magnetohydrodynamic (MHD) pressure drop, with implications on heat and mass transfer. The current density in the fluid can be reduced by decoupling electrically the liquid metal from the well-conducting blanket structure by thin-walled flow channel inserts (FCI) as proposed for the DCLL concept [1]. Sandwich-type FCIs consist of an electrically insulating ceramic layer protected from all sides by thin sheets of steel to avoid direct PbLi-ceramic contact. The steel layers are thin to provide sufficient Ohmic resistance for limiting currents. In order to support design activities for DCLL blankets, 3D numerical MHD simulations have been performed for pressure driven MHD flows in a DCLL blanket duct (Fig. 1). The liquid metal enters the blanket structure radially at the bottom, flows in a poloidal channel along the plasma-facing first wall, makes a U-turn at the top and flows down in a second poloidal duct before it leaves the module through the back plate. Fig. 1 shows the design of a DCLL blanket module and the simplified model used for the numerical simulations.

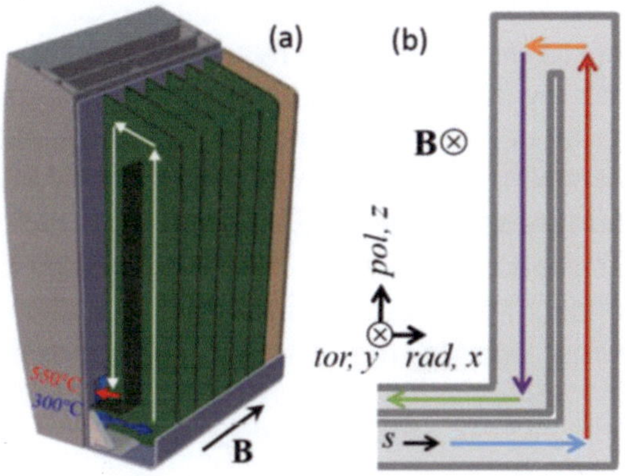

Fig. 1 Design of a DCLL blanket module [2] (a) and principle sketch of flow path (b).

The physical parameters characterizing the MHD flow are the Hartmann number and the Reynolds number,

$$Ha = BL\sqrt{\frac{\sigma}{\rho\nu}}, \quad Re = \frac{u_0 L}{\nu},$$

where L and u_0 denote typical length scale and velocity. Ha^2 und Re stand for the ratio of electromagnetic to viscous forces and inertia to viscous forces, respectively. In the present work L=0.0808m stands for half the duct size measured in toroidal direction and u_0 is the average velocity at the inlet. The conductance of the FCI has been taken into account by numerically resolving the thin conducting sheets with thickness t_w=0.5mm [2]. The computational mesh with $5.7 \cdot 10^6$ cells consists of structured hexahedral cells with non-uniform spacing for higher resolution of boundary layers and conducting sheets of the FCIs. The lengths of entrance and exit ducts have been extended compared with the original design in order to impose fully established entrance conditions and reasonable convective parallel outflow. The numerical method uses an extension of the hydrodynamic open source code Open-FOAM for MHD applications [3].

Figure 2 shows contours of electric potential and velocity streamlines in the model geometry at the mid-plane of the duct for moderate *Ha*=500 and fusion-relevant high *Ha*=8000. 3D effects are present near bends and U-turn, but the flow approaches quite rapidly fully established conditions in the poloidal channels. Streamlines follow approximately lines of constant electric potential. Flow separation and recirculation, visible for *Ha*=500 behind the bends, are suppressed for higher magnetic fields. Despite the high Reynolds number *Re*=10932 the flow remains stable and laminar for the parameters considered.

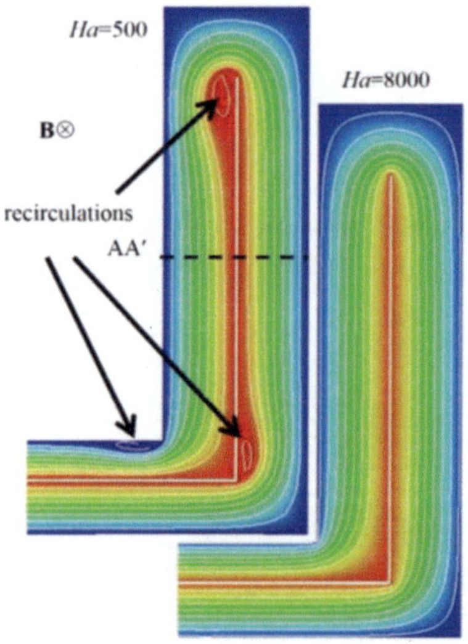

Fig. 2 Contours of potential and streamlines in a DCLL blanket for u0=2cm/s, for Ha=500 and Ha=8000.

The pressure distribution along a centerline of the blanket module is shown in Figure 3 for *Ha*=8000 and two velocities u_0=1cm/s and u_0=2cm/s. Bends and U-turns do not substantially contribute to the total pressure drop, because the flow turns in radial-poloidal planes perpendicular to **B** [4]. Immediately after passing 3D elements, the pressure gradient practically coincides with that of an asymptotic analysis for fully developed flows.

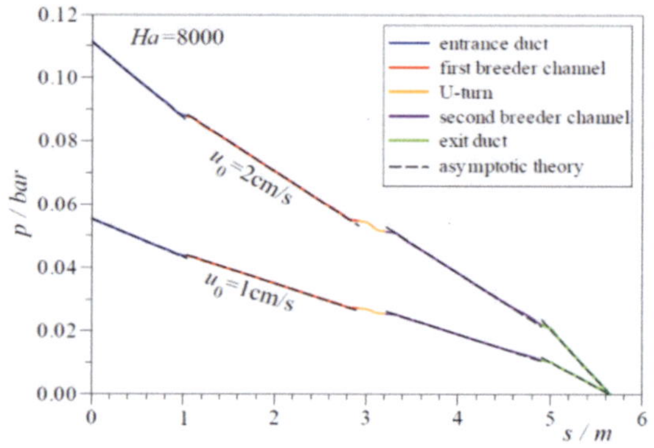

Fig. 3 Pressure as a function of a coordinate *s* along the flow path in the center of the channels.

Simulations for various Hartmann numbers 500 ≤ *Ha* ≤ 8000 and flow rates show that the total pressure drop scales as $\Delta p \sim u_0 B^2$, the typical scaling for inertialess MHD flows in strong magnetic fields. Present results, published in [5], should be considered as a first step for predicting liquid metal flow in DCLL blankets. In order to complement the present work, future simulations will take into account in addition inclined fields, heat transfer and strong buoyancy effects for which the laminar flow might become unstable or even turbulent.

Experimental study of liquid metal MHD flows near gaps between FCIs

For reduction of pressure drop in DCLL blankets, electrically insulating flow channel inserts (FCIs) are foreseen for electrically decoupling the liquid metal flow from the well-conducting channel walls. For fabrication reasons, gaps between inserts are unavoidable. Gaps interrupt the electrical insulation, thus providing unwanted local short circuit for electric currents. For experimental investigations of 3D effects at junctions of FCIs, a test section has been manufactured and experiments are currently performed in the MEKKA facility [6] using NaK as model fluid. The experimental study shows the benefits of FCIs for pressure drop reduction in fully developed flows as also predicted by theoretical analyses [7] and quantifies the deterioration of pressure drop reduction by the presence of uninsulated gaps between FCIs.

The test section used for experimental investigations has a circular cross-section. Details of the geometry and dimensions, including the radius R_{FCI} of the inner steel layer of the FCI, are shown in Figure 4. The thick-walled pipe has an inner radius R=L that is used as length scale for the problem. The thickness t_{FCI}=0.5mm of the protection sheets has been suggested e.g. in [2]. More details about the instrumentation and first results can be found in [8].

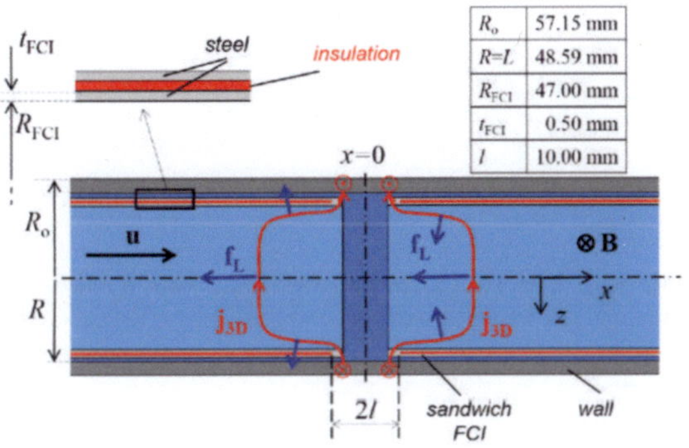

R_o	57.15 mm
R=L	48.59 mm
R_{FCI}	47.00 mm
t_{FCI}	0.50 mm
l	10.00 mm

Fig. 4 Geometry at the junction between two FCIs.

The intensity of the flow and strength of the magnetic field are quantified by the non-dimensional Reynolds and Hartmann numbers Re and Ha. The influence of wall conductivity σ_w on MHD pipe flow is described using the wall conductance parameter c according to [9], which evaluates for the present problem and material properties at 50°C to

$$c = \frac{\sigma_w}{\sigma} \frac{R_0^2 - R^2}{R_0^2 + R^2} = 0.0727$$

for the thick-walled pipe and to c_{FCI}=0.00476 for the thin inner conducting sheet of the FCI. For strong magnetic fields, the nondimensional pressure gradients in fully established flow become $\partial p/\partial x$=0.0678 for the pipe without FCI and to $\partial p/\partial x$=0.00508 for an infinitely long FCI, respectively [8].

For investigations of 3D effects near the non-insulated region at x=0, the gap between FCIs has been positioned in the center of the magnet. Pressure differences with respect to the position x=0 have been measured along the axis and normalized by $\sigma u_0 B^2 L$. In the following, x stands for the axial position, scaled by L. The insulations in both FCIs extend from x=±0.2 up to x=±12.25 as illustrated in the subplot in Fig. 5Figure 5.

Results displayed in Figure 5 for Ha=3000 show the great benefit from using FCIs by comparing the theoretical curves for pipe flow and assumed perfect FCI. The theory for fully developed MHD flows predicts a pressure drop reduction factor of 0.0678/0.00508=13.3. At some distance from the gap, here for $|x|>1$, the measured pressure gradients in the FCIs are close to these predictions as indicated in the figure. Experiments confirm that a pressure drop reduction close to theoretical values is achievable with the used sandwich-type inserts. At larger distance from the center, the pressure gradient increases. This is caused by additional 3D effects, since for $|x|>8.8$ the magnetic field is not uniform anymore and the insulation ends at $|x|$=12.24. 3D effects at entrance and exit of FCIs will be investigated in future experiments. At x=0 some additional pressure drop Δp_{3D}=0.12 is present due to leakage currents across the gap into the well-conducting pipe wall. There seems to be no difference in non-dimensional pressure drop near the gap for the Reynolds numbers considered, i.e. those flows are apparently inertialess.

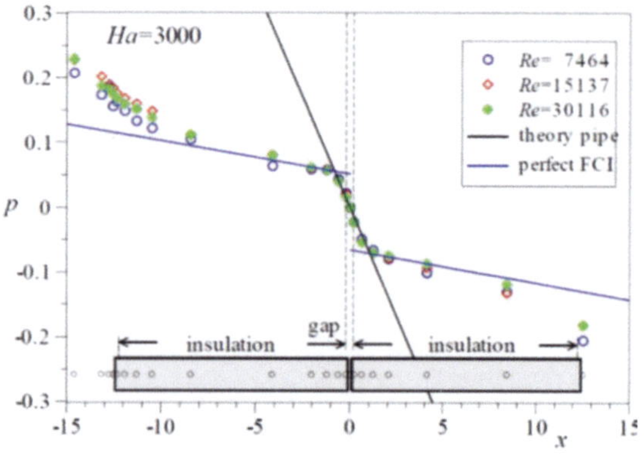

Fig. 5 MHD flow at the junction between two FCIs. Pressure along the axis for *Ha*= 3000 and different *Re*. Positions of FCIs and pressure taps are illustrated in the sketch.

According to these results, sandwich-type FCIs may reduce pressure drop of fully developed MHD flow by more than one order of magnitude compared to that in a thick-walled pipe. The additional non-dimensional pressure drop at a gap between two FCIs Δp_{3D}=0.12 seems not very high. Nevertheless, one should keep in mind that this pressure drop caused by one single non-insulated gap corresponds to that of a fully developed flow in an FCI over a length Δl_{3D}=23.6.

Fabrication of thin-walled flow channel inserts

Manufacturing of straight rectangular sandwich-type FCIs for DCLL blankets have been achieved by standard technologies such as coating, bending and welding [10]. However, it turned out that these techniques cannot be applied to circular FCIs as required in the experiments described above. The fabrication of sandwich-type FCIs for geometries deviating from rectangular cross section requires a different method, which has been developed. The process starts with a thin-walled pipe as inner sheet. The latter is coated on the outer surface with an insulating ceramic layer leaving blank small areas at both ends and a narrow stripe along the pipe (see Figure 6). A second metal sheet is wrapped around the coated pipe and laser-welded along all edges. The small longitudinal non-coated stipe foreseen in the experiment for measurement purposes (pressure taps) would not be required for FCIs in blanket ducts. After successful fabrication, the FCIs have been exposed to high temperatures above 300°C. Afterwards all welds have been checked by a dye penetrant test, which did not show any crack or defect. One prototype FCI has been cut into pieces for microscopic inspection of weld samples. The welds in all samples show no cracks or pores and a difference with respect to the original material could not be seen. More details about the fabrication of this circular FCI can be found in [11]. The method proposed here (coating, wrapping, welding), proven to be efficient for fabrication of circular FCIs, should apply as well for other types of cross section shapes that do not have sharp corners.

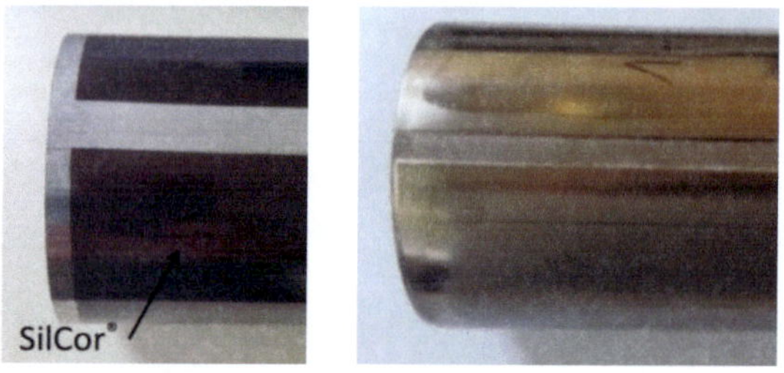

Fig. 6 Pipe coated with SilCor®SiC (left), wrapped and laser-welded (right).

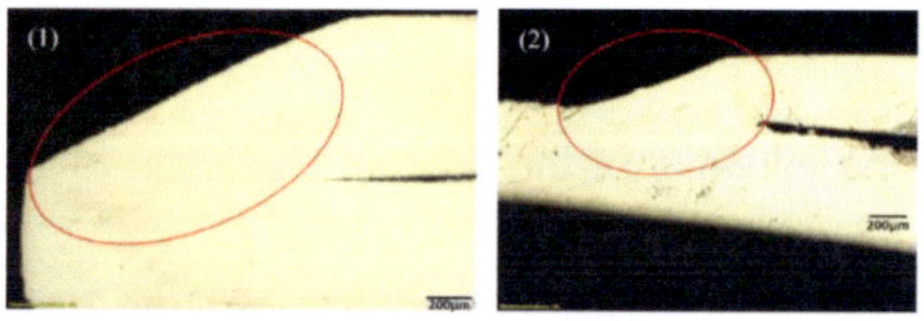

Fig. 7 Samples of polished specimens: (1) edge of the orbital weld, (2) longitudinal weld.

The fabrication of insulating 3D components like complex shaped FCIs for DCLL blankets is more challenging. Selective laser melting (SLM) is an additive manufacturing technique, in which a high-power-density laser melts metallic powders gradually in layers to create solid 3D structures. This technique has been tested for fabrication of complex thin and double-walled insulating 3D FCIs. Preliminary studies for SLM-produced components have been made to check material properties and to fabricate hybrid elements for blankets such as stiffening plates with cooling channels [12], [13].

The FCIs produced by SLM have been designed as a downscaled model of a liquid metal channel in the DCLL blanket (Figure 8Fig. 8). The 3D shape of the mockup follows the liquid metal flow path inside the blanket. Objective is to show the general feasibility of fabricating complex 3D double-walled FCI structures with very thin walls by SLM. Walls should be as thin as currently achievable with SLM and produced from reduced activation martensitic Cr-Mo steel Eurofer97. The fact that Eurofer97 is not a standard material for SLM and walls should be as thin as possible, poses a major challenge for the manufacturer. The first mockup therefore is considered as a test to gain experience about practical limitations of present-day SLM technique. The FCI model has been designed with a height of 110mm. It consists of a double-walled structure with 0.8mm wall thickness each and an insulating gap of 0.8mm. The inner and outer walls are initially separated from each other by spacers at both openings to ensure exact positioning of both walls. After completion of the model by SLM, the gap between the layers has been filled with ceramic powder (ø0.3mm), the spacers have been removed and the gaps were closed by non-conducting caps (Fig. 9).

Inspection of this first FCI mockup shows that some of the walls have certain deformations. Electrical resistance measurements confirm undesired contacts. Examinations by x-ray computed tomography (CT) show the positions, where inner and outer walls might be in contact and deteriorate the electrical isolation.

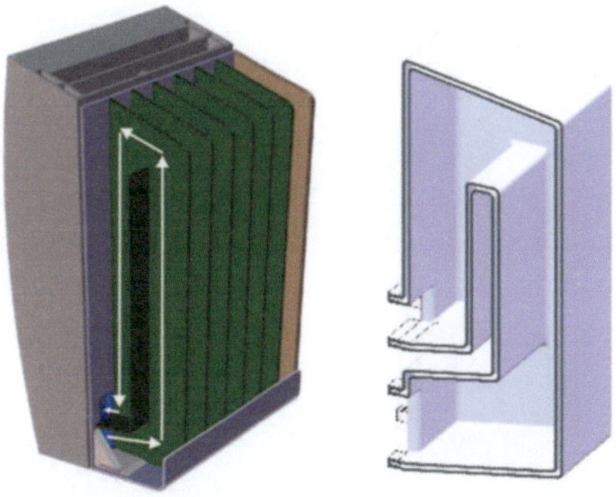

Fig. 8: Present design of DCLL blanket (left) [2] and CAD model of the downscaled double-walled SLM-fabricated FCI mockup (right).

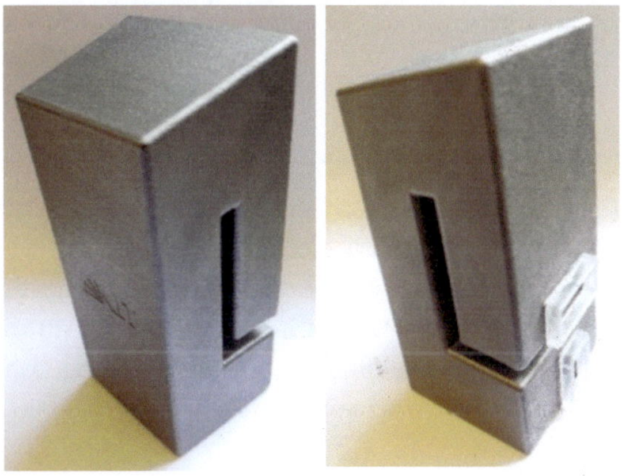

Fig. 9: FCI filled-up with ceramic powder and closed by caps.

After optimization of the SLM process, a second mockup has been fabricated with slightly increased thickness of 1mm for each wall and an insulating gap of 1.4mm. It has been observed and confirmed by CT scans, that walls are now much less deformed (see Figure 10). Electric resistance measurements confirm perfect electrical insulation and therewith ensured full functionality of the mockup for electrically decoupling of the liquid metal flow from the well conducting blanket structure. For more details see [14].

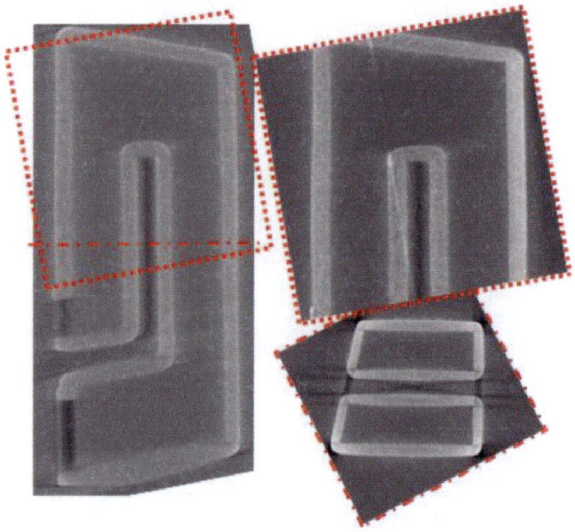

Fig. 10: CT scan of the improved FCI mockup (1mm wall thickness, 1.4 mm nominal gap): Longitudinal cut (left), detail (right, top), cross section (right, bottom). *(by Andreas Meier, KIT-IAM)*

The present work should be considered as proof of suitability for production of 3D sandwich-type insulating elements for liquid metal blankets, based on available technology for SLM. For fabrication of full-scale components, SLM machines would require new approaches [13] and upscaling in dimensions.

Design of a test section for analysis of magneto-convection in water-cooled lead lithium blankets

Another liquid metal blanket concept considered for a DEMO reactor is the water-cooled lead lithium (WCLL) breeding blanket [15]. The liquid metal alloy lead lithium PbLi is used as breeder, neutron multiplier and as heat carrier. Pressurized water flows through a large number of cooling pipes immersed in the liquid metal and cools the walls and breeding zone (see Figure 11). In order to get first insight in the distribution of the PbLi flow resulting from the combined interaction of electromagnetic forces, buoyancy and pressure, preliminary numerical and experimental studies of buoyant MHD flows are foreseen. A simplified model geometry will be used for these studies to improve the understanding of magneto-convective flows and to demonstrate the feasibility of the proposed WCLL DEMO design. First results for flows in vertical channels with internal obstacles have been reported in [16].

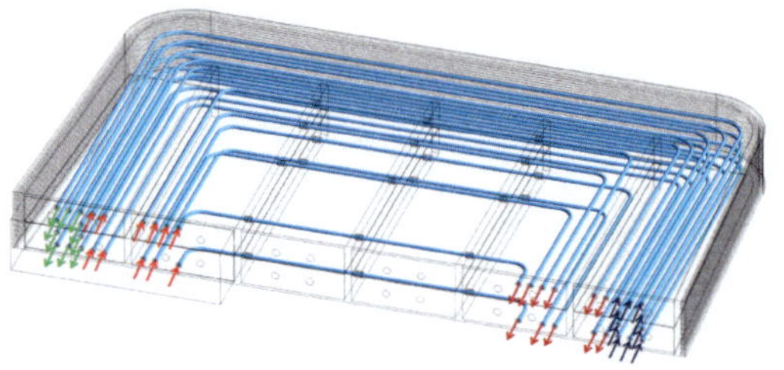

Fig. 11 WCLL blanket (ENEA [14]), equatorial section in a blanket module with cooling pipes.

Figure 12 shows the CAD design of the experimental test section. It consists of a liquid-metal-filled box, in which a heated pipe simulates neutron heating while a cooled pipe removes the heat from the liquid metal. The magneto-convective flow is driven by buoyancy, when heat is exchanged between liquid metal and the cooling and heating pipes. The dimensions of the box have been chosen such that the experiment fits into the gap of the dipole magnet available in the MEKKA laboratory.

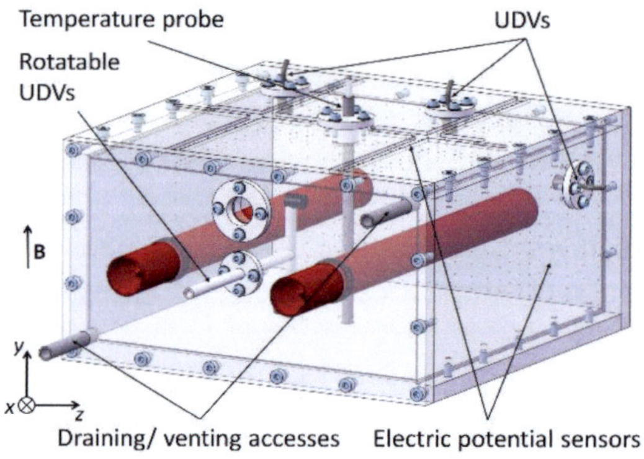

Fig. 12 Design of the experimental test section.

The pipes are supplied with copper cores for achieving as good as possible isothermal conditions. Thermocouples at different axial and circumferential positions monitor the temperature during the experiments near the inlet, outlet and at the center of the pipes (see Figure 13).

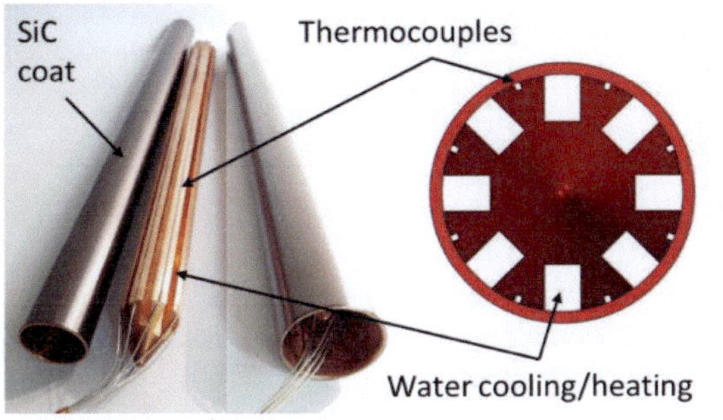

Fig. 13 Copper pipe and core with inserted thermocouples (left), front view of design (right).

In order to support the design of the present model experiment with some theoretical data, numerical simulations have been performed, in which the driving temperature difference between heated and cooled pipe has been set to ΔT=T2-T1=40K. The strength of the magnetic field is measured in terms of the Hartmann number *Ha*. All walls are electrically insulating and adiabatic. Some results of numerical simulations are shown in Figure 14. The figures in the left column show contours of fluid temperature. Flow streamlines indicate that there are convective cells present in the center between the two pipes and the maximum velocity occurs near the walls of the tubes. According to these results, suitable positions for measuring temperature are along magnetic field lines. Interesting velocity data is expected in layers tangent to the pipes, and electric potential differences might be seen only on the end walls of the box, where electric currents have to close in thin viscous layers.

The temperature distribution in the center of the test section is measured by a temperature probe with 11 thermocouples equidistantly distributed over the height (Figure 15). The temperature on the walls of the box is monitored in more than 50 positions in order to get an overview of the temperature distribution and to estimate the amount of transferred heat.

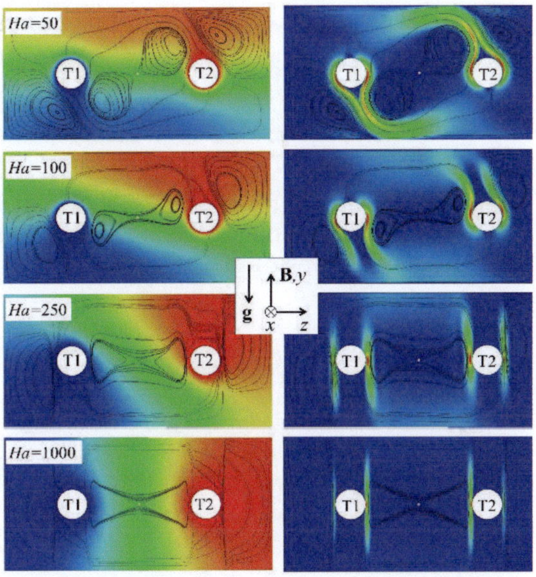

Fig. 14 Magneto-convective flow for ΔT=T2-T1=40K. Contours of temperature (left) and magnitude of velocity (right).

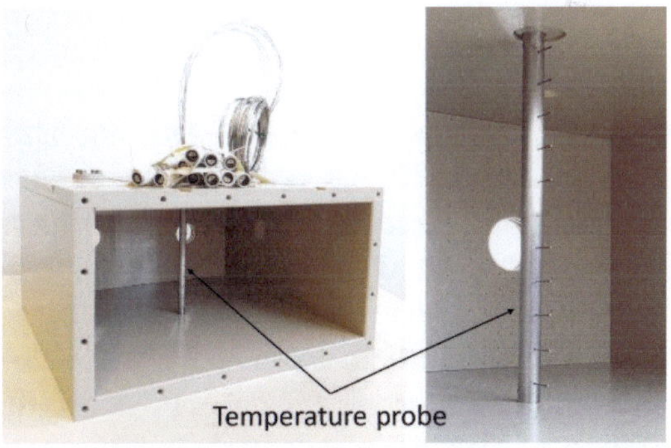

Fig. 15 Temperature probe in the test section.

The electric potential will be measured on the fluid-solid interface by a large number of copper electrodes that penetrate specific walls of the test section. Most interesting walls, the end walls, are instrumented with a dense population of electrodes.

Numerical simulations show that the highest velocities occur in thin layers tangent to the heating/cooling pipes. It is planned to record velocity data from those layers by using Ultrasound Doppler Velocimetry (UDV) [17]. A principle sketch showing the positions of the sensors and the propagation zones of the ultrasonic pulses is displayed in Figure 16. Two sensors will give information about the flow in the tangent layers and one in the cores. For measuring velocity along other directions, a movable probe with two UDV sensors is foreseen. The probe is mounted on a rotatable rod and can be turned in *yz* plane.

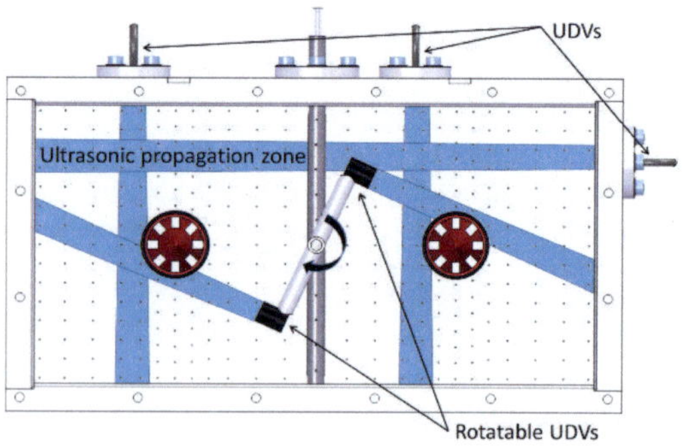

Fig. 16 Sketch showing positions of UDV sensors on the walls and rotatable probe. Propagation zones of ultrasonic pulses in blue.

Before the test section will be filled with liquid metal, first experiments are performed using water as a test fluid. These preliminary tests are foreseen to show full functionality of the instrumentation, and they will give first results for the hydrodynamic limit, when no magnetic field is present. Water allows optical access and application of Particle Image Velocimetry (PIV) for flow visualization. For this purpose, end walls and one sidewall are replaced by transparent material (see Figure 17).

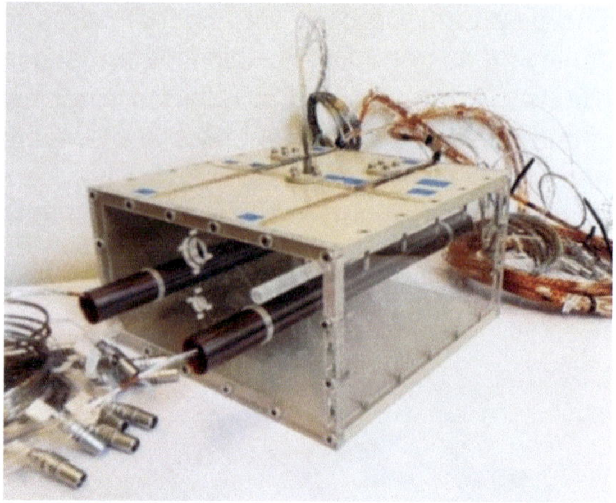

Fig. 17 WCLL mockup for preliminary experiments with water: front, side and back walls made of Plexiglas.

The test section is now instrumented and ready for measuring simultaneously temperature, potential and velocity distribution. It is foreseen to perform experiments in 2019 for various values of the applied magnetic field and driving temperature differences, i.e. for various Hartmann and Grashof numbers, respectively.

Further work

In addition to the topics described above, the MHD group at IKET KIT contributed in reporting period 2017-2018 with scientific papers to the development of helium cooled lead lithium (HCLL) blankets [18], [19], [20] [21], [22], to sensor development [23], and to fundamental studies of MHD flows [24], [25], [26], [27], [28], [29], [30].

References

[1] Rapisarda, D., Fernandez, I.; Palermo, I.; Urgorri, F.; Maqueda, L.; Alonso, D.; Melichar, T.; Frýbrt, O.; Vála, L.; Gonzalez, M.; Norajitra, P.; Neuberger H. and Ibarra, A.; "Status of the engineering activities carried out on

the European DCLL," Fusion Engineering and Design, 2017.

[2] Fernández-Berceruelo, I.; Gonzalez, M.; Palermo, I.; Urgorri F. and Rapisarda, D.; "Large-scale behavior of sandwich-like FCI components within the EU-DCLL operational conditions," Fusion Engineering and Design, p. in print, 2018.

[3] Mistrangelo, C. and Bühler, L.; "Development of a numerical tool to simulate magnetohydrodynamic interactions of liquid metals with strong applied magnetic fields," Fusion Science and Technology, vol. 60, no. 2, pp. 798-803, 2011.

[4] Molokov, S.; "Liquid metal flows in manifolds and expansions of insulating rectangular ducts in the plane perpendicular to a strong magnetic field," 1994.

[5] Klüber, V.; Mistrangelo, C. and Bühler, L.; "Numerical simulations of 3D magneto-hydrodynamic flows for dual-coolant lead lithium blankets," in Proceedings of the 30th Symposium on Fusion Technology SOFT, Giardini Naxos, Italy, September 16-21, 2018, 2018.

[6] Barleon, L. ; Mack, K.-J. and Stieglitz, R.; "The MEKKA-facility a flexible tool to investigate MHD-flow phenomena," 1996.

[7] Bühler, L. and Mistrangelo, C.; "Pressure drop and velocity changes in MHD pipe flows due to a local interruption of the insulation," Fusion Engineering and Design, vol. 127, pp. 185-191, 2018.

[8] Bühler, L.; Brinkmann, H.-J. and Koehly, C.; "Experimental study of liquid metal magnetohydrodynamic flows near gaps bet-ween flow channel inserts," in Proceedings of the 30th Symposium on Fusion Technology SOFT, Giardini Naxos, Italy, September 16-21, 2018, 2018.

[9] Miyazaki, K.; Konishi, K. and Inoue, S.; "MHD pressure prop of liquid metal flow in circular duct under variable transverse magnetic field," Journal of Nuclear Science and Technology, vol. 28, no. 2, pp. 159-161, 1991.

[10] Norajitra, P.; Basuki, W.W.; Gonzalez, M.; Rapisarda, D.; Rohde, M. and Spatafora, L.; "Development of sandwich flow channel inserts for an EU DEMO dual coolant blanket concept," Fusion Science & Technology, vol. 68, no. 3, pp. 501-506, 2015.

[11] Koehly, C. and Bühler, L.; "Fabrication issues of sandwich-like flow channel inserts for circular pipes," Fusion Science and Technology, vol. 72, pp. 660-666, 2017.

[12] Neuberger, H.; Rey, J.; Hees, M.; Materna-Morris, E.; Bolich, D.; Aktaa, J.; Meier, A.; Fischer, S.; Schorle, C.; Fuhrmann, U.; Heger, R.; Dlouhy, I.; Stratil, L. and Kloetzer, B.; "Selective laser sintering as manufacturing process for the realization of complex nuclear fusion and high heat flux components," Fusion Science and Technology, vol. 72, no. 4, pp. 667-672, 2017.

[13] Neuberger, H.; Rey, J.; Hernandez, F.; Ruck, S.; Arbeiter, F.; Rieth, M.; Koehly, C.; Stratil, L.; Niewoehner, R. and Felde, A.; "Evaluation of conservative and innovative manufacturing routes for gas cooled Test Blanket Module and Breeder Blanket First Walls," in Proceedings of the 30th Symposium on Fusion Technology SOFT, Giardini Naxos, Italy, September 16-21, 2018, 2018.

[14] Koehly, C.; Neuberger, H. and Bühler, L.; "Fabrication of thin-walled fusion blanket components like flow channel inserts by selective laser manufacturing," in Proceedings of the 30th Symposium on Fusion Technology SOFT, Giardini Naxos, Italy, September 16-21, 2018, 2018.

[15] Romanelli, F.; Barabaschi, P.; Borba, D.; Federici, G.; Horton, L.; Neu, R.; Stork, D. and Zohm, H.; "Fusion electricity: A roadmap to the realisation of fusion energy," EFDA, 2012.

[16] Bühler, L. and Mistrangelo, C.; "MHD flow and heat transfer in model geometries for WCLL blankets," Fusion Engineering and Design, vol. 122, pp. 919-923, 2017.

[17] Brito, D.; Nataf, H.-C.; Cardin, P.; Aubert, J. and Masson, J.-P.; "Ultrasonic Doppler Velocimetry in Liquid Gallium," Experiments in Fluids, vol. 31, pp. 653-663, 2001.

[18] Bühler, L.; Mistrangelo, C.; Brinkmann, H.-J. and Koehly, C.; "Pressure distribution in MHD flows in an experimental test-section for a HCLL blanket," Fusion Engineering and Design, vol. 127, pp. 168-172, 2018.

[19] Mistrangelo, C. and Büher, L.; "Determination of multichannel MHD velocity profiles from wall-potential measurements and numerical simulations," Fusion Engineering and Design, vol. 130, pp. 137-141, 2018.

[20] Mistrangelo, C.; Bühler, L. and Brinkmann, H.-J.; "Experimental investigation of MHD pressure losses in a mock-up of a liquid metal blanket," Nuclear Fusion, vol. 58, p. 036012, 2018.

[21] Mistrangelo, C.; Bühler, L.; Koehly, C. and Brinkmann, H.-J.; "Influence of modifications of HCLL blanket design on MHD pressure losses," Fusion Engineering and Design, vol. 124, pp. 948-952, 2017.

[22] Mistrangelo, C. and Bühler, L.; "Magnetohydrodynamic flows in liquid metal blankets for fusion reactors," Proceedings in Applied Mathematics and Mechanics, vol. 17, pp. 115-118, 2017.

[23] Bühler, L.; Aiello, G.; Bendotti, S.; Koehly, C.; Mistrangelo, C. and Galabert, J.; "Development of combined temperature – electric potential sensors," Fusion Engineering and Design, vol. 136, pp. 7-11, 2018.

[24] Arlt, T.; "Analyse des Stabilitätsver-haltensmagnetohydrodynamischer Kanalströmungen," PhD Thesis, Karlsruhe Institute of Technology, 2018.

[25] Krasnov, D.; Bandaru, V.; Bühler, L. and Boeck, T.; "Instabilities and turbulence in magnetohydrodynamic duct flows," in Proceedings of the NIC Symposium 2018. 22-23.02.2018, Jülich, Germany, vol. 49, K. Binder, M. Müller and A. Trautmann, Eds., Jülich, Germany, John-von-Neumann Institute for Computing, 2018, pp. 389-396.

[26] Arlt, T.; Priede, J. and Bühler, L.; "Influence of thin finitely conducting walls on the linear stability of magnetohydrodynamic flows," Magnetohydrodynamics, vol. 53, no. 1, pp. 35-44, 2017.

[27] Arlt, T.; Priede, J. and Bühler, L.; "The effect of finite-conductivity Hartmann walls on the linear stability of Hunt's flow," Journal of Fluid Mechanics, vol. 822, pp. 880-891, 2017.

[28] Chowdhury, V.; "Experimentelle Untersuchungen von magnetohydrodynamisch induzierten Instabilitäten," PhD Thesis, Karlsruhe Institute of Technology, 2017.

[29] Krasnov, D.; Boeck, T. and Bühler, L.; "Turbulent and transitional sidewall jets in magnetohydrodynamic channels with a homogeneous magnetic field," Proceedings in Applied Mathematics and Mechanics, vol. 17, pp. 111-114, 2017.

[30] Mistrangelo, C. and Bühler, L.; "Instabilities in electrically driven rotating MHD layers," IOP Conference Series: Materials Science and Engineering, vol. 228, p. 012004, 2017.

Group: Transmutation

Analytical Modeling of the Emergency Draining Tank for a Molten Salt Reactor

Mattia Massone, Shisheng Wang, Andrei Rineiski

Abstract

The Molten Salt Fast Reactor is a reactor concept developed by the European Union based on a liquid fuel salt circulating through the reactor core. A peculiar emergency system, which takes advantage of the liquid fuel state, is represented by a tank located underneath the core, where the fuel can be passively drained and cooled; its geometry ensures that the fuel remains in subcritical conditions.

In the framework of the SAMOFAR project, a design for the Emergency Draining Tank has been proposed: the tank shall be equipped with vertical cooling elements, arranged in a hexagonal grid; the liquid fuel salt, which heats up due to decay heat, will fill the gaps between the elements.

In this work, analytical methods (Green's functions and orthogonal decomposition) are employed to study the transient heat transfer associated with the proposed design and to perform a preliminary dimensioning of the system, such that overheating is avoided in any moment of the transient and the fuel salt is kept in a liquid state and in safe conditions for a long time. The models are constituted by multilayer monodimensional slabs and cylinders, with a pure heat conduction model. The assessment of the available grace time and preliminary considerations about fuel salt freezing and its influence on the system effectiveness are also included.

Key Words: MSFR, reactor safety, transient heat conduction, emergency draining, Green's function

Introduction

The Molten Salt Reactor concept, originally developed at the Oak Ridge Laboratories [1] is currently under study by many public and private actors. Among them, the European Union is developing the Molten Salt Fast Reactor Concept (MSFR) with the SAMOFAR project [2], in the framework of the HORIZON-2020 program.

The used fuel is a binary salt, composed of lithium and actinides fluorides. At the operating temperature (~725 °C), the fuel is in liquid state and continuously flows through the core and the intermediate heat exchangers. The geometrical configuration ensures that criticality is achieved only in the core region, where the liquid heats up due to the fission reaction; a fraction of the nominal power (3 GWth) is, however, released outside of the core in form of decay heat [3], [4].

Taking advantage of the liquid state of the fuel, an Emergency Draining System has been included in the design: in case the fuel overheats it can be passively relocated (through freeze plugs) in the Emergency Draining Tank (EDT), where it is kept in safe conditions. The geometry of the tank is specifically designed to keep the fuel salt in subcritical conditions, but the decay heat has to be somehow evacuated to ensure long-term safety.

Gérardin et al. [5] investigated in the past the design of the EDT from the neutronic point of view, to ensure the system subcriticality, and performed preliminary calculations on the heat transfer. The present work aims at studying different options from the heat transfer point of view with the intention to suggest a design solution that assures the safety of the system and meets the EDT specific objectives.

Model Description

Geometry

The proposed EDT is a cavity filled with hexagonal cooling assemblies, arranged in a hexagonal grid. The drained fuel is transferred from the core to the EDT through a funnel-shaped draining shaft [6] ending above the central assembly, which does not have cooling capability but helps achieving an equal distribution of the fuel salt in the tank [4], [5].

Each assembly, as shown in Fig. 1, is constituted by a water pipe surrounded by a layer of an inert material enclosed in a steel casing; the fuel fills the gaps between the assemblies [[4]][[5]].

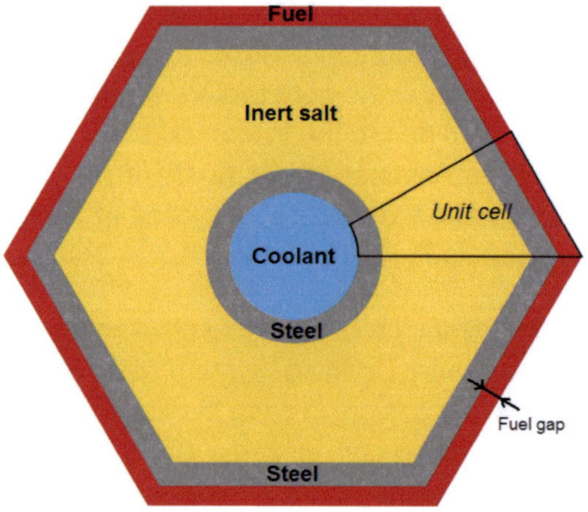

Fig. 1. EDT cooling assembly scheme and model unit cell

This work will consider different options for the inert and fuel salt thickness; the 3-cm-steel layer, instead, will not be changed, as it is considered a needed safety measure to keep a robust structure that prevents the fuel salt from mixing with the inert material. The influence of the steel components on the temperature profile is, however, very limited, due to their high thermal conductivity and low thermal capacity.

It is clearly possible (and advisable) modeling only the unit cell of the assembly, i.e. the rectangular triangle representing 1/12 of the whole assembly and depicted in Fig. 1.

Design Objectives

The EDT aims at ensuring the coolability of the fuel salt that, as a result of the accumulation of fission products and transuranics due to fission, has an internal heat generation due to decay heat; such time-dependent generation is provided by a fitted curve [[4]], based on the reactor isotopic composition.

The design should be capable of preventing the fuel salt from boiling. In addition, as the structural integrity of the tank should be kept, a safety limit should be set to avoid any deterioration of the steel mechanical properties. The maximum admissible temperature is set to 1200°C, which is considered a safe margin before the steel mechanical properties are compromised [4]. Being such value lower than fuel boiling point, above 2000 K [7], the temperature limit is entirely defined by the steel.

The EDT is intended to be used, in principle, only in case of emergency [4]; anyhow, one cannot exclude that spurious discharge can occur or that, in some cases, the system can be employed in other circumstances. In all cases, it is not acceptable that the fuel salt becomes useless once drained; this means that the draining operation should be reversible. Such point requires basically two conditions: the fuel salt in the EDT does not form a homogeneous mixture with other materials; the salt is kept in a liquid state for a long time, so that it can be pumped back into the core as soon as the cause of the draining has been removed.

In order to increase the safety level provided by the design, this should be able to provide additional grace time [5], i.e. the intervention of the operators should not be required immediately to ensure the fuel coolability. The study of the coolant system, or whether it should be operated in natural or forced circulation, is not within the scope of this paper; however it is important to estimate the maximum time allowed before the coolant flow is available, compatible with the compliance with the safety limits of the EDT.

Main Assumptions

Convection

The convective motion is not considered in the model. The establishment of natural circulation cannot be excluded, as the cooling on the steel side combined with the internal heating and symmetry condition on the other side will lead to a temperature gradient. Nevertheless, the space between the two plates is very small and the flow can be disturbed by local fuel solidification and by the initial conditions of the draining; also, the circulation scheme could be more complicated than expected, involving more unit cells in a systemic circulation. However, as it is considered that the natural convection can only improve the heat transfer by favoring remixing of the fluid, the exclusion of the phenomenon can only make the calculation of the system maximum temperature more conservative; on the contrary, for the same reason, the fuel temperature at the interface with the cooling assembly might be higher than calculated. Hence the match with the safety limit will be checked with the maximum system temperature rather than the maximum steel one.

More precisely, with regards to its motion, the fuel salt is considered to be completely at rest, as if it were a solid. This conflicts with the chaotic movement expected by the fuel salt during the draining, which will take some time before it settles down. Anyhow, the advantages provided by the exclusion of the convective motion (namely, the model is more conservative) applies also in this case.

Initial temperature

We hypothesize that the flushing is instantaneous as the reactor stops operating: no fuel energy is lost or gained during the draining and the initial temperature is the average operating temperature in the core, i.e. 725 °C [4]. Similarly, the cooling assemblies are not preheated, and they are initially at room temperature, assumed to be 20 °C.

This constitutes an approximation because, actually, the draining can take a few minutes to be completed, and it has an influence on the fuel initial temperature and on the generated heat flux. The initial temperature, in fact should reflect the time required for the freeze plugs of the core to open, the heating of the collector, the energy released through the draining shaft to the surrounding air... Similarly, the decay heat generated starts dropping as soon as the system is not critical anymore, i.e. before the fuel reaches the EDT.

Material properties

The temperature dependence of the material properties is neglected; the introduction of temperature-variable properties, in fact, would change radically the problem characterization, which would become non-linear [8]. The analytical solution of a non-linear problem is much more complicated and not always feasible; in particular, in a non-linear system the superposition of effects, which is at the base of the Green's function approach, cannot be applied. For similar reason, the latent heat could also not be taken into account.

However, as the internal (coolant tube) and external (casing) steel walls are expected to have very different temperatures, their constant properties (related to the same kind of steel) are sampled for the temperature ranges which are most close to the expected conditions (as shown in Table I). Being the dimensioning of the tank still in a preliminary phase, specific steel has not been selected; the 316 stainless steel [[9]] has then been used. This choice has hardly an effect on the results of our study, as the thermal transfer is dominated by the other materials, which have much lower thermal conductivities and higher heat capacity.

The material inside the cooling assembly has the function to conduct the heat to the cooling tube during its functioning and provide thermal inertia to the system during the grace time. As a large amount of heat can be stored as fusion latent heat, the ideal material should be solid at room temperature, but melt at a temperature below the maximum allowable temperature, i.e. 1200 °C, assumed as safety limit for the steel. Hence, FLiNaK [10] has been chosen.

Analytical Formulation

The basic relation to describe the heat conduction is the diffusion equation [8]

$$(\boldsymbol{x})\nabla^2 T(\boldsymbol{x},t) + \frac{\alpha(x)}{\lambda(x)}\dot{q}(\boldsymbol{x},t) = \frac{\partial T(x,t)}{\partial t}, \quad (1)$$

being α the thermal diffusivity, λ the thermal conductivity and $\dot{q}$ the internal volumetric heat generation rate of the material.

The space dependence of the material properties can be removed by applying the equation separately for each of the M homogenous regions R_i

$$\alpha_i \nabla^2 T_i(\boldsymbol{x},t) + \frac{\alpha_i}{\lambda_i}\dot{q}_i(t) = \frac{\partial T_i(x,t)}{\partial t} \qquad i = 1..M \quad (2)$$

and imposing the temperature and heat flux continuity on the boundary surfaces S_i (perfect thermal contact) as boundary conditions (BCs)

$$T_i(\boldsymbol{x},t) = T_{i+1}(\boldsymbol{x},t) \quad \forall \boldsymbol{x} \in S_i, i \in (0,M) \quad (3)$$

$$\lambda_i \frac{\partial T_i(x,t)}{\partial n} = \lambda_{i+1}\frac{\partial T_{i+1}(x,t)}{\partial n} \quad \forall \boldsymbol{x} \in S_i, i \in (0,M). \quad (4)$$

The problem is closed with the initial conditions defined in §2.3.2 and imposing a symmetry condition on the fuel side and a fixed temperature condition on the water side

$$\frac{\partial T_M(x,t)}{\partial n} = 0 \qquad \forall \boldsymbol{x} \in S_M \quad (5)$$

$$T_1(\boldsymbol{x},t) = T_\infty \qquad \forall \boldsymbol{x} \in S_0. \quad (6)$$

This latter condition implies that the water is considered as a perfect heat sink, kept at constant temperature.

The solution of the presented problem [8] is a convolution of the Green's functions G_{ij}

$$T_i(\boldsymbol{x},t) = \sum_{j=1}^{M}\Big[\int_{R_i} G_{ij}(\boldsymbol{x},t|\boldsymbol{\xi},\tau)\big|_{\tau=0}\cdot \tilde{T}_j(\boldsymbol{\xi})\, d\boldsymbol{\xi} + \frac{\alpha_i}{\lambda_i}\int_0^t \dot{q}_j(\tau)\int_{R_i} G_{ij}(\boldsymbol{x},t|\boldsymbol{\xi},\tau)\, d\boldsymbol{\xi}\, d\tau\Big]. \quad (7)$$

While the first term considers the initial conditions, which can be regarded as an instantaneous impulse, the second one takes into account the effect of the internal heat generation, composed by the contribution of all the impulses from the initial time, appropriately weighted depending on the elapsed time. It is

Table I. Materials thermal properties

	Fuel salt [[7]]	Inert salt [[10]] (FLiNaK)	Steel [9]	
			Coolant tube (~100 °C)	Casing (~700 °C)
$\boldsymbol{\lambda}\ (\boldsymbol{W}\cdot\boldsymbol{m}^{-1}\cdot\boldsymbol{s}^{-1})$	1.7	0.905	15.1	24.5
$\boldsymbol{\alpha}\ (\boldsymbol{m}^2\cdot\boldsymbol{s}^{-1})$	3.70 10^{-7}	2.40 10^{-7}	3.72 10^{-6}	5.39 10^{-6}

important noticing that the Green's function depends on two regions: the function represents the way the region *j* affects the temperature distribution in *i*.

Assessment of the Mathematical Model

The height of each cooling assembly is much larger (in the order of meters) than the transversal dimensions (dm): hence the reduction from a 3D to a 2D geometry is straightforward and has a negligible effect on the results accuracy, at least as far as convection is excluded.

The 2D section described in §2.1, instead can be easily approximated by a multilayer trapezoid but, even if the X direction is larger than the Y one, the difference is not large enough to assume the discrepancy between 2D and 1D models would be negligible. This poses an issue from the mathematical point of view: in fact, none of the 11 orthogonal coordinate systems listed by Morse and Feshbach [11] as separable for the Helmholtz equation can be used to represent the trapezoid-like geometry. This means that the eigenfunctions of the Helmholtz equation, i.e. the homogeneous heat conduction problem in space, are not a product of functions depending each one on a single coordinate. In other words, the separation of the space coordinates is not a viable path, and eigenfunctions depending on both coordinates at the same time should be looked for.

The problem of eigenfunctions in non-separable space domains is an extremely interesting topic, still discussed in applied mathematics [12], [13], but the presentation and application of these methods goes beyond the scope of this paper, aiming at a preliminary dimensioning of the EDT elements. Hence, we will approximate the system as either a slab or a cylinder in the X direction (i.e. on the vertical sections of the assembly passing through the symmetry center). The discrepancy with the actual 2D temperature profile is expected to be very small in the vicinity of the hexagon apothem and increase while approaching the hexagon circumradius.

The Green's Function

The separation of variables can anyhow be applied to the space and time parameters and the solution, once applied the initial and BCs, yields the Green's function [8] to be used in (7)

$$G_{ij}(x,t|\xi,\tau) = \frac{\lambda_j}{\alpha_j}\sum_{n=1}^{\infty}\left\{\begin{matrix}\left[\sum_{j=1}^{M}\frac{\lambda_j}{\alpha_j}\int_{x_{j-1}}^{x_j}\psi_{j,n}^2(\xi)\,d\xi\right]^{-1} \\ e^{-\beta_n^2(t-\tau)}\psi_{i,n}(x)\psi_{j,n}(\xi)\end{matrix}\right\} \tag{8}$$

The normalization integral depends on the adopted geometry, and includes an additional *x* term for the cylindrical geometry. In (8) appear the eigenfunctions ψ_n, which are solutions of the eigenvalue problem associated to the homogeneous diffusion equation (1) and the BCs: the shape of the solutions depends on the system geometry

$$\psi_{i,n}^{\text{slab}}(x) = A_{i,n}\sin\left(\frac{\beta_n}{\sqrt{\alpha_i}}x\right) + C_{i,n}\cos\left(\frac{\beta_n}{\sqrt{\alpha_i}}x\right) \tag{9}$$

$$\psi_{i,n}^{\text{cylinder}}(x) = A_{i,n}\,\mathrm{J}_0\left(\frac{\beta_n}{\sqrt{\alpha_i}}x\right) + C_{i,n}\,\mathrm{Y}_0\left(\frac{\beta_n}{\sqrt{\alpha_i}}x\right), \tag{10}$$

while the values of the coefficients A_i, C_i and of the eigenvalues $\boldsymbol{\beta_n}$ are determined using the BCs.

The Eigenvalues

By imposing the BCs (3)-(6) one obtains a system of 2*M* equations and 2*M* unknowns ($A_{i,n}$ and $C_{i,n}$); in order to exclude the trivial solution and get the needed eigenvalues and eigenvectors, an arbitrary unknown is set to a non-zero value. The eigenvalues of the problem are those for which the determinant of the associated matrix is null [8].

Fig. 2 shows the value of the associated matrix determinant depending on β_n; the zeros of the functions can be found using a numerical method (e.g. the bisection method). In principle, for slab geometry the function is periodic, as it is a composition of periodic functions, but the period is often so large that it includes more β_n than actually needed to have a good representation; hence, its determination has no practical use.

Results

The procedure described above has been coded into the MATLAB software. The following results are calculated using such software and truncate the series in (8) after the first 1000 β_n.

Fuel layer

As a first step in the assessment of the EDT element dimension, we test the effectiveness of the system in the transient depending on the thickness of the fuel layer. In order to have the highest thermal inertia, the initial thickness of the inert salt is set to 15 cm, and will be later reduced. As a first approximation, a slab geometry is considered.

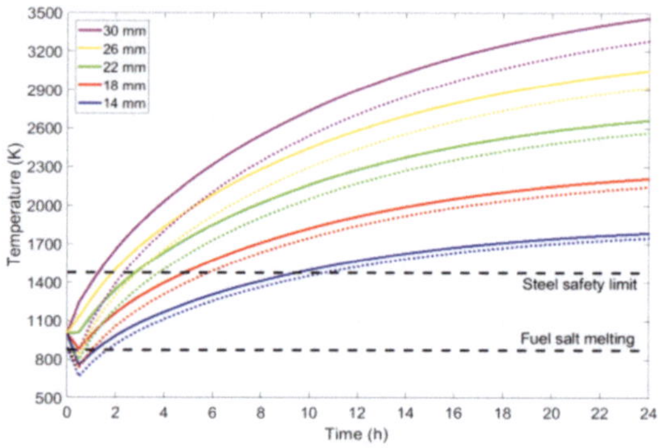

Fig. 3. Maximum (solid line) and minimum (dotted line) fuel temperature along the transient depending on the fuel layer thickness.

As shown in Fig. 3 the thickness of the fuel salt has a huge impact on the maximum temperature of the system; in fact, the larger the amount of fuel, the larger the heat generated in the system that has to be removed by each cooling assembly. With all tested fuel gaps the temperature keeps increasing even after one day; in addition, none of the systems is able to stay below the steel safety limit. However, it is not considered a valid option reducing the gap below 28 mm due to the high risk of blockage in the draining procedure. We will then proceed with such fuel gap, reducing the inert salt thickness instead.

Inert salt layer

In order to comply with the limit, we consider reducing the inert salt thickness, which hinders

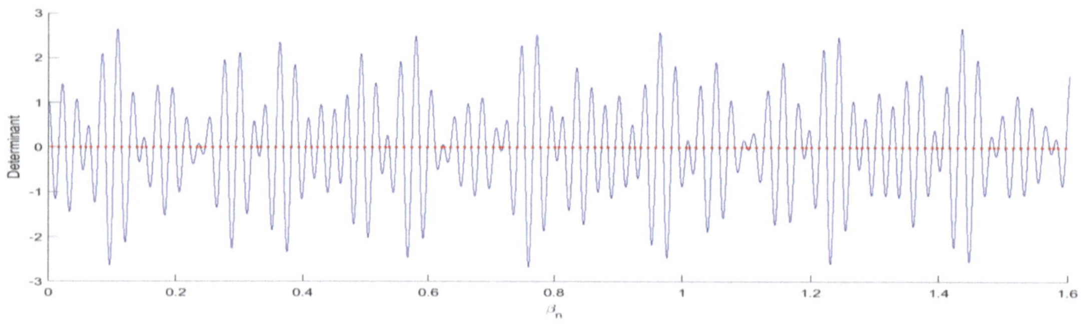

Fig. 2. Eigenvalue search for a slab case.

the heat transfer due to its low thermal conductivity, especially in the later phases of the transient, when the thermal capacity is saturated.

One can see in Fig. 4 that the temperature trend of the different options start to diverge after a couple of hours, i.e. when the thermal capacity of the 7 cm inert salt layer approaches saturation; in a temperature profile, like Fig. 5, a measure of the remaining available heat capacity is given by the curvature of the profile (its second derivative): for a non-heat-generating material, a positive curvature indicates that some energy can still be stored, while a negative one denotes an excess of energy stored, e.g. due to the reduction of the temperature of the heated side.

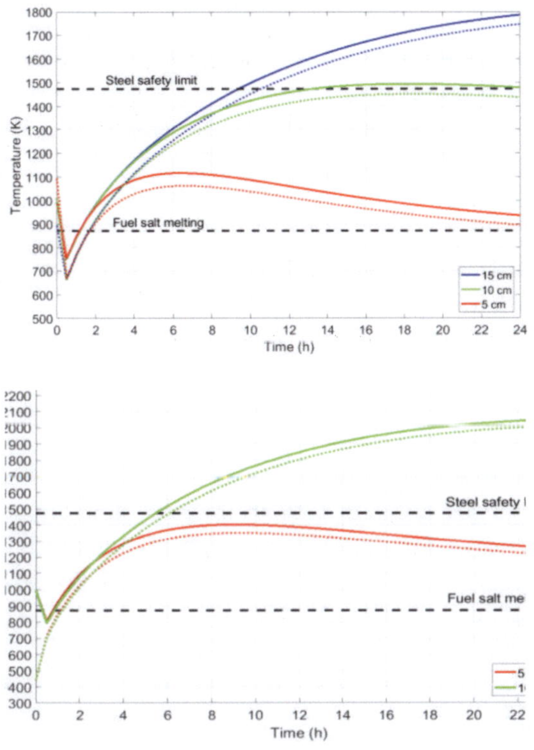

Fig. 4. Maximum (solid line) and minimum (dotted line) fuel temperature along the transient depending on the inert salt layer thickness (top: slab; bottom: cylindrical geometry).

The two figures show also that with the 5 cm configuration, a maximum temperature is achieved after 4-8 h (depending on the geometry); after this moment, the heat generated is lower than the energy removed by the cooling. As such maximum temperature is below the steel safety limit, we can conclude that the system is able to keep the fuel in a long-lasting safe condition.

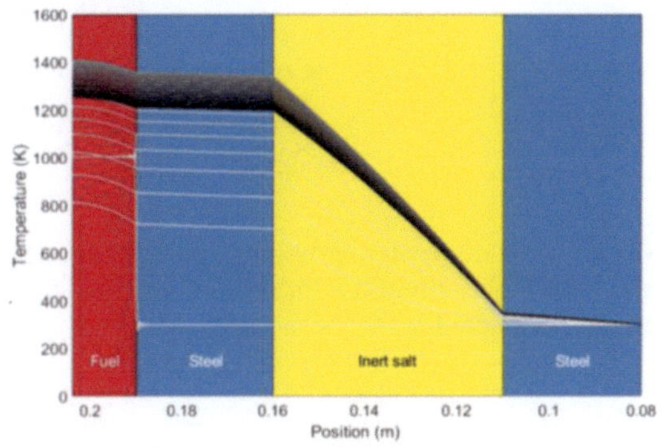

Fig. 5. Temperature profile (cylinder) of the 14/50 mm EDT element between 0 and 24 h, with intervals of 30 minutes (lines color is darker as time progresses).

<u>*Grace time*</u>

The large thermal inertia provided by the FLiNaK layer allows using the EDT as a completely passive system for some time before switching on the coolant flow.

Fig. 6 shows the effect of the delayed activation of the cooling on the maximum temperature of the fuel salt, i.e. the main safety issue considered in this paper. It is clear that a delay of up to 2 hours has almost no influence on the fuel temperature, and one can also wait for 3 hours with no concern on the temperature limits. In case the grace time is extended to 4 hours, instead, the temperature increases enough to slightly exceed the steel safety limit.

In all cases, the plot shows that there is about 30 minutes delay between the cooling activation and the first visible effect on the fuel maximum temperature.

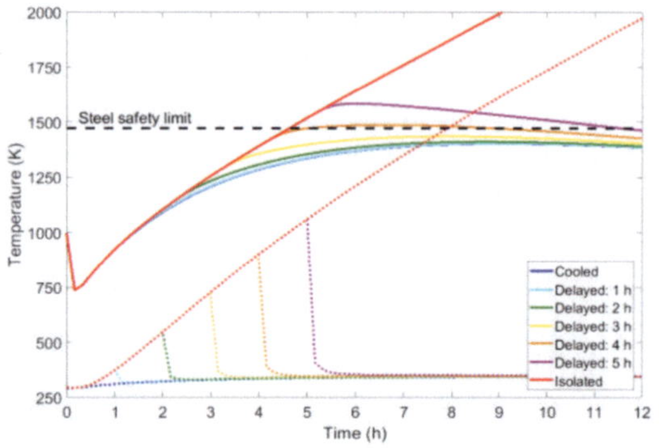

Fig. 6. Fuel maximum temperature (solid lines) and inert salt minimum temperature (dotted line) along the transient.

Fuel salt solidification

The assessment of the solidification behavior of the frozen salt represents an important matter for the EDT design. In the draining phase, a quick solidification should be avoided, as it could block the gaps and create an obstacle to a uniform and complete discharge. In addition, the fuel salt liquid state is required to ensure reversibility of the draining operation.

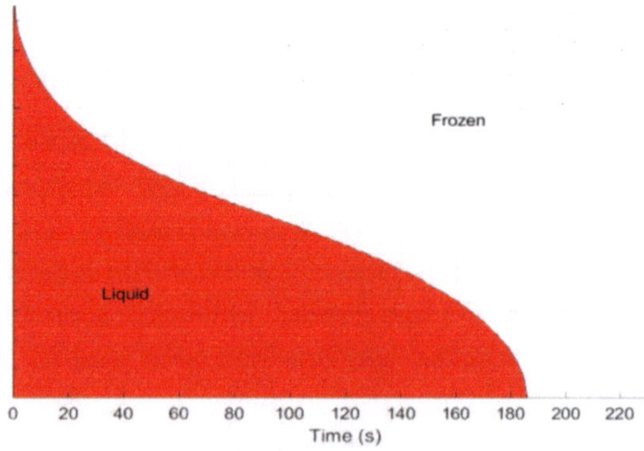

Fig. 7. Available fuel gap for salt flow in the first minutes after draining.

The freezing of part of the fuel on the cooling assembly walls cannot be completely avoided if the EDT is initially at room temperature. Fig. 7 one shows that the fuel temperature drops below the freezing point (872 K) after 3 minutes; hence, if the draining were completed within such time, the salt could be distributed uniformly in the tank without any blockage risks. On the contrary, if more time were required, a detailed transient 2D analysis, including also the chance of stratification of the solid fuel in the Z-direction, would be needed to assess the expected distribution of the salt and be sure that the draining can be completed safely. Actually, the time required to freeze the inter-assembly gap is longer, as the simulation does not take into account the effect of the latent heat, that should be removed too before freezing actually occurs

Reversibility is guaranteed by the fact that complete remelting of the fuel salt, due to the decay heat, is achieved within 1 h 30', as shown in Fig. 4. It has been already shown that the maximum fuel temperature reaches a peak between 8 and 12 h after the discharge. Once the generated heat is lower than what the coolant is capable of removing, an adequate handling of the coolant flow allows to keep under control the energy evacuated and so the temperature of fuel, which hence can be kept above the melting point for a long time. This ensures the reversibility of the system in the long term.

However, it should be taken into account that, due to the solidification/remelting behavior, any draining process forces the reactor to stop for a couple of hours, being such time needed for the fuel salt to melt again. Also, one should consider the effect of the thermal expansion in freezing/melting processes to exclude the possibility of structural damage.

The solidification issue is one of the reasons why using more conductive materials (e.g. lead) is not advised: the heat transfer through the layers, in fact, is so fast that a remelting of the fuel salt could not be guaranteed. In addition, lead has a thermal inertia much lower than the FLiNaK, hence making difficult providing grace time, as discussed in §4.3.

Conclusions

This study aims at analyzing the performance of the emergency draining tank of the molten salt reactor based on the temperature profiles calculated analytically using a Green's functions and orthogonal decomposition in a 1D geometry. The analytical approach proved to be very suitable for this kind of study, as it can be easily applied to simplified geometries and provide results which are accurate enough but can be obtained in very short time and with limited computational resources. It is considered of interest for future studies the extension into a 2D geometry, employing the state-of-the-art knowledge regarding the computation of non-separable spaces eigenfunctions.

The results show that it is possible to find a layout of the proposed tank design that is able to keep the drained fuel temperature below 1200 °C, representing the structural integrity limit. The same layout allows a grace time up to 3 hours after the draining, during which no cooling at all is required. Once passed a few hours (between 8 and 12) from the draining, the generated heat is lower than the evacuated heat, assuring the safety of the design; from this point on, a careful tuning of the removed heat flux can be used to keep the fuel salt in a liquid state for a very long time, ensuring the reversibility of the draining operation. It is however not possible to avoid the solidification of the fuel during the draining in a containment which is kept at room temperature; fuel gaps take at least 3 minutes to become completely blocked, but this time does not take into account the effect of the latent heat, which should instead be considered in future studies. The solidified fuel, however, melts again completely within 90 minutes; this however means that any discharge results in a reactor stop, to allow the fuel to melt again.

The proposed design proved to be able to comply with the objectives required in very conservative conditions. Future analyses should start from this point, focusing on removing the assumptions made for this analytical modeling (e.g. by including the temperature-dependence of the material properties and assessing the convective motion) and so refining the design.

Acknowledgements

The authors thank the anonymous reviewer for his/her accurate review and insightful comments.

The authors would like to acknowledge the financial support by the Euratom research and training programme 2014-2018, under the Grant Agreement Number 661891 – SAMOFAR.

Partners

We thank Paul Servell, Grenoble INP – Phelma, for his contribution to EDT studies, including code development and preliminary calculations.

References

[1] Nuttin et al.; "Potential of thorium molten salt reactors: detailed calculations and concept evolution with a view to large scale energy production", Progress in Nuclear Energy, 46, pp. 77-99 (2005).

[2] "A Paradigm Shift in Reactor Safety with the Molten Salt Fast Reactor", https://cordis.europa.eu/project/rcn/196909_en.html (2017).

[3] Wang, S., et al.: "A Passive Decay Heat Removal System for Emergency Draining Tanks of Molten Salt Reactors", *Proceedings of the Int. Topical Meeting NURETH-17*, Xi'an (China), 3-8 September 2017.

[4] Allibert, M., et al.: "D1.1 Description of initial reference design and identification of safety aspects", *A Paradigm Shift in Nuclear*

Reactor Safety with the Molten Salt Fast Reactor (grant agreement number: 661891) – SAMOFAR (2017).

[5] Gérardin, D., et al.; "Design Evolutions of the Molten Salt Fast Reactor", *Proceedings of the Int. Conf. FR17*, Yekaterinburg (Russian Federation), 26-29 June 2017.

[6] Wang, S., et al.; "Analytical Investigation of the Draining System for a Molten Salt Fast Reactor", *Proceedings of the Int. Topical Meeting NUTHOS-11*, Gyeongju (Korea), 9-13 October 2016.

[7] Beneš, O., et al.; "D3.2 Physico-Chemical properties of the MSFR fuel salt", *Evaluation and Viability of Liquid Fuel Fast Reactor System (grant agreement number: 249696) – EVOL* (2013).

[8] Necati Özişik, M.; *Heat Conduction – Second Edition*, Wiley, New York, USA (1993).

[9] Waltar, V.; Reynolds, A.; *Fast Breeder Reactors*, Pergamon Press, New York, USA (1980).

[10] Beneš, O.; Konings, R.J.M.; "Thermodynamic properties and phase diagrams of fluoride salts for nuclear applications", *Journal of Fluorine Chemistry*, **130**, pp. 22-29 (2009).

[11] Morse, P. M.; Feshbach, H.; *Methods of Theoretical Physics, Part I*, McGraw-Hill, New York, USA (1953).

[12] Betcke, T.; Trefethen, L. ; "Reviving the Method of Particular Solutions", *SIAM Rev.*, **47(3)**, pp 469-491 (2005).

[13] Descloux, J.; Tolley, M.; "An accurate algorithm for computing the eigenvalues of a polygonal membrane", *Comput. Methods Appl. Mech. Engrg.*, **39**, pp. 37-53 (1983)

Group: Severe Accident Research

Analysis of Severe Accidents

Xiaoyang Gaus-Liu, Giancarlo Albrecht, Thomas Cron, Beatrix Fluhrer, Jerzy Foit, Hiroshi Madakoro, René Stängle, Mike Vervoortz, Thomas Wenz

Introduction

The research activities on reactor severe accident in LWRs at the group IKET-UNA and since September 2018 named IKET-SAR from 2017 to 2018 are focused on the in- and ex-vessel core melt behavior under diverse core melt accident scenarios and reactor types. These include the melt pool formation and retention in reactor lower head, melt jet dispersion in a flooded containment cavity, and corium concrete interaction and corium coolability in a containment cavity.

Following experimental works were carried out in the two-years period:

- LIVE3D and LIVE2D: series of tests on heat transfer of homogenous and stratified melt pool
- DISCO: two Fuel Coolant Interaction (FCI) tests in reactor pit filled with water
- MOCKA: two tests on ex-vessel molten corium and Concrete Interaction (MCCI) process with basaltic concrete.

Besides the experimental activities, calculation work has been performed on the coupling of a reactor analysis code and a lower head thermal analysis solver.

Some of the important results are described in following:

Heat transfer in a stratified melt pool

General

In the course of melt pool formation in LWR lower head, the melt could be separated into several layers because of the non-miscibility of melt components. The most concerned scenario is a formation of a light metallic layer atop of an oxide pool. In this case, the decay power in the oxide pool is transferred upwards to the metallic layer. The metallic layer is then heated up and constitutes a very high thermal load on the reactor wall at this part. Unclear is how the thickness of the metallic layer influences the heat flux on the sidewall and the role of the interlayer crust on the heat transfer between the two layers. The stratified melt pool tests are high demanding on the selection of non-miscibility simulant liquids in a possible operational temperature.

Experiment

In the frame of EU H2020 IVMR project, the LIVE2D vessel was upgraded for the conduction of the melt stratification tests. The upgrade measure includes the installation of transparent front side on the test vessel, for direct visual observation, more temperature measurements in the upper-layer region, local and environmental extraction system of possible oil vapor and automatic pool temperature control and power shutdown.

The simulant materials were determined after a series of laboratory examination. A thermal oil with high stability up to 240°C is used for the upper layer and the eutectic nitrate salt

$NaNO_3$-KNO_3 with operational liquid state temperature 224 - 350 °C simulates the oxidic layer with power dissipation.

As the first on its kind in Europe and worldwide, SAR experimental team has performed successfully a series of tests with separated melt layers and were able to demonstrate the complex interaction among the power input in the lower layer, the upper layer thickness, the boundary cooling conditions and the structure of the interlayer crust. In Figure 1 the formation of the interlayer crust at lower power input (900W) and the meltdown on the interlayer crust at higher power input of 1400W are shown

Fig. 1. Two-layer melt pool with the power input of 900W (top) and 1400W (bottom)

The movement of turbulence in the two layers can be captured with a real-time video camera; whereas the long-time process of crust formation and melt down in the whole view is recorded in time lapse. Temperature distribution in the two melt layers and on the vessel wall provide additional information of the turbulent pattern in the melt and the heat flux on the vessel wall.

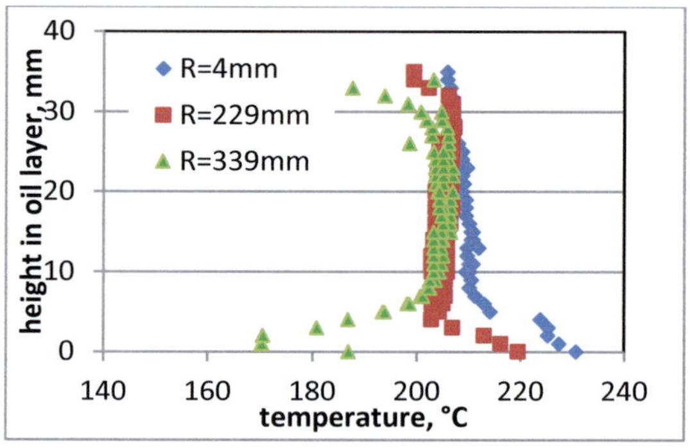

Fig. 2. Melt temperature distribution at three radial positions in a 35mm upper layer

Ex-vessel fuel coolant interaction (Ex-FCI) experiment in the DISCO test facility

Tests performed in DISCO facility investigate the phenomena during an ex-vessel FCI. A high-temperature iron–alumina melt is ejected under high pressure from RPV bottom to reactor cavity of a certain reactor design.

The major objective of these experiments was to determine the shape and size of the melt jet/steam in a transparent cavity. The influence of the geometry of the cavity and the openings to the containment sub-containments and dome on the pressure escalation and particle distribution in reactor is another objective of the study.

The test data include (i) high-speed video data to visualize the melt jet and water interaction, (ii) melt/steam volume in interaction zone, (iii) pressure evolution in reactor cavity and other parts of a containment, (iv) temperatures in RPV/RCS vessel, cavity and containment vessel, (v) characterization of melt debris, (vii) break opening characteristics and dispersal of water and melt at the end of interaction.

The experiments were carried out in the DISCO-H facility which was used before for Direct Containment Heating (DCH) experiments and the FCI experiments mentioned above. In Figure 3 at the left, the DISCO-H facility and at the right, the transparent cavity are shown.

Fig. 3: Left: containment pressure vessel and internal structures of RPV/RCS and cavity. right: the cavity (water pool), a cuboid vessel with glass windows at three sides

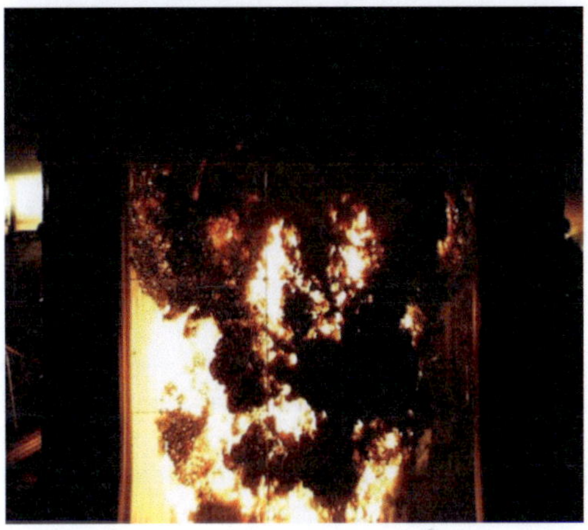

Fig. 4 Vizualisazion of the interaction zone by means of a high-speed video recording. Top: t=26ms, bottom: t=295ms

After ignition of the thermite the break of the wire installed close to the brass plug was detected about 10 s later. This time is defined as t = 0 s. The 2200°C hot melt was ejected by nitrogen at 1.0 MPa into water at 50°C. The maximum pressure in the water vessel was 0.32 MPa after 0.6 seconds. The pressure in the containment increased from 0.1 MPa to 0.17 MPa, Figure 5. No steam explosion occurred. Most of the melt debris was found in the water vessel and had a median diameter of 2 mm, Figure 6.

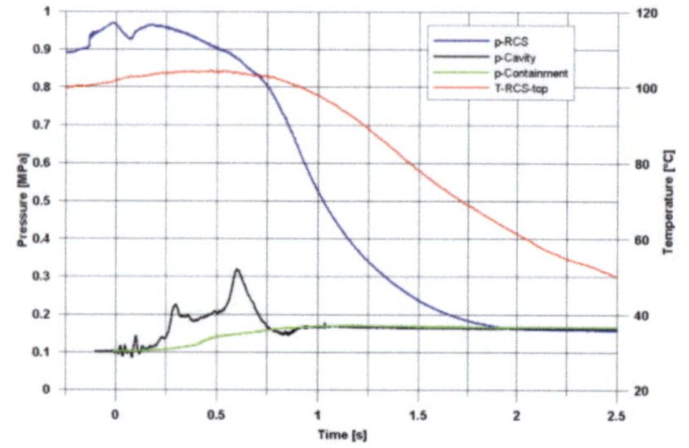

Fig. 5: Pressure in reactor cooling system, cavity and containment

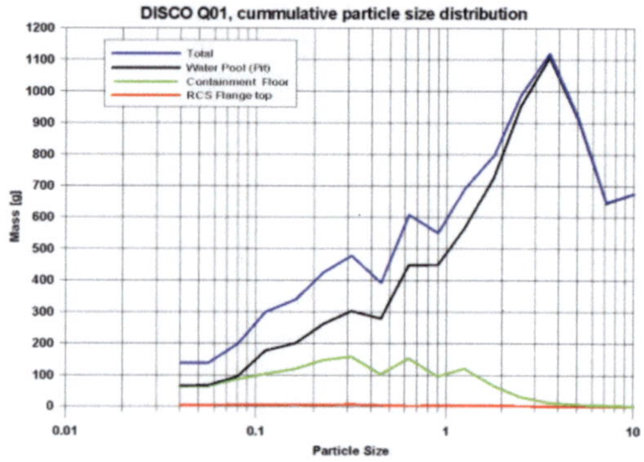

Fig. 6: Mass distribution of the collected debris

MOCKA-SSM experiments on basaltic concrete with and without rebars

The large-scale MOCKA (Metal Oxide Concrete Interaction in KArlsruhe) experiments at KIT investigated the interaction of a simulant oxide (Al_2O_3, ZrO_2, CaO) and metal melt (Fe) in a stratified configuration with different types of concrete. The program was focused on assessing the influence of a typical 12 wt% reinforcement in the concrete on the erosion behavior. The developed method of using the heat of chemical reactions to simulate the decay heat allows to obtain a rather prototypic heating of both melt phases and to study the interaction of the melt with a reinforced concrete.

A series of large-scale MOCKA experiments studied the interaction of the melt with siliceous and LCS concrete. Different erosion behaviour between siliceous and limestone concrete were found in several 2D MOCKA experiments. A highly pronounced lateral ablation was observed in all MOCKA tests using LCS concrete with and without rebars. The corresponding lateral/axial ratios of the ablation depth are approximately 3 In the MOCKA experiments with siliceous concrete, the overall downward erosion by the metal melt was of the same order as the sideward one. In addition, the lateral erosion in the overlaid oxide melt region was about the same as in the metal melt region. Fig 7 shows MOCKA experiment performance.

Fig. 7: MOCKA test facility.

To study the erosion behaviour of basaltic concrete, typical for Swedish BWRs, two experiments have been performed in the MOCKA test facility within the EU project of FP7 SAFEST. MOCKA-SSM1 with and MOCKA-SSM2 without reinforcement in the basaltic concrete, Figure 8.. For both MOCKA-SSM tests, concrete crucibles with an inner diameter of 250 mm are used. Both the sidewall and basement are instrumented with Type K thermocouple assemblies to approximately determine the position of the progressing melt front into the concrete as well as tungsten-rhenium thermocouples for measuring the melt temperatures. A total of 51 thermocouples are implemented.

The basaltic concrete does not contain a concrete plasticizer, consequently, the fluidity of the concrete was considerably reduced. This in turn led to difficulties in distributing the concrete within the rebar structure. Therefore, some cavities were formed in the concrete.

The preparation of the crucible without rebars was completely unproblematic.

Fig. 8: MOCKA-SSM1 rebar structure

The melt is directly generated in the test crucible by a thermite reaction of 80 kg thermite powder and 30 kg CaO. The resulting melt consists of 42 kg Fe, overlaid by 68 kg oxide melt (initially 56 wt.% Al2O3, 44 wt.% CaO). The collapsed height of the metal melt is about 13 cm and that of the oxide melt 50 cm. The initial melt temperature at start of interaction is approximately 1800 °C. The CaO admixture lowers the solidus temperature and the viscosity of the melt. The resulting solidus temperature of approx. 1360 °C is sufficiently low to prevent a formation of an initial crust at the oxide/concrete interface.

After the completion of the thermite burn thermite and Zr was added to the melt within approximately 40 minutes in the experiments under consideration. The heat generated by the thermite reaction and the exothermal oxidation reaction of Zr is mainly deposited in the oxide phase. Due to density-driven phase segregation the metal melt at the bottom of the crucible is fed by the enthalpy of the steel which is generated in the oxide phase by the thermite reaction of the added thermite. Taking the added Zr and thermite into account, approximately 75 % of the heating power is deposited in the oxide phase and 25 % in the metal melt.

A highly irregular final shape of the ablation was observed in the MOCKA-SSM2 without reinforcement, Figure 9. This irregular shape was caused by a strong concrete spallation process, which resulted in simultaneous splash out of a substantial amount of oxide melt. The post-test cavity erosion profile of MOCKA-SSM1 with reinforced concrete showed a more regular shape of the erosion front, because the reinforcement seems to stabilize the concrete against the spallation process. In the tests with siliceous and LCS concrete such a behaviour during MCCI was not observed. However, more MCCI experiments on basaltic concrete are needed to confirm the observed phenomena. If the strong spallation is an inherent property of the basaltic concrete, then it must be considered as an issue for nuclear safety assessment.

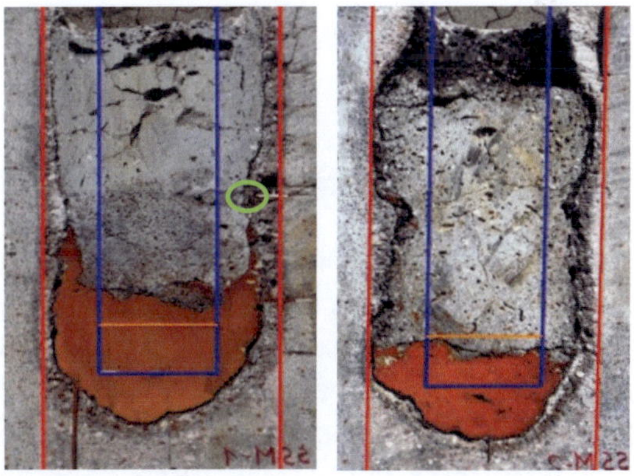

Fig. 9. Section of the MOCKA-SSM1 (left) and MOCKA-SSM2 (right) concrete crucible with an indication of the initial size of the crucible cavity (blue line). The orange line indicates the initial height (13 cm) of the metal melt and the red line marks the outer surface of the basaltic cylindrical crucible.

Coupling of a reactor analysis code and a lower head thermal analysis solver

Due to the recent high interest on IVR, development of de-tailed thermal and structural analysis tool, which can be used in a core-melt severe accident, is inevitable. Although RELAP/SCDAPSIM is a reactor analysis code, originally developed for US NRC, which is still widely used for severe accident analysis, the modeling of the lower head is rather simple, considering only a homogeneous pool. PECM/S, a thermal structural analysis solver for the RPV lower head, has a capability of predicting molten pool heat transfer as well as detailed mechanical behavior including creep, plasticity and material damage. The boundary condition, however, needs to be given manually and thus the application of the stand-alone PECM/S to reactor analyses is limited. By coupling these codes, the strength of both codes can be fully utilized. Coupled analysis is realized through a message passing interface, OpenMPI. The validation simulations have been performed using LIVE test series and the calculation results are compared not only with the measured values but also with the results of stand-alone RELAP/SCDAPSIM simulations

Analysis method

- RELAP/SCDAPSIM

The RELAP/SCDAPSIM code, designed to predict the behavior of reactor systems during normal and accident conditions, is being developed as part of the international SCDAP Development and Training Program (SDTP). RELAP/SCDAPSIM consists of three separate modules: RELAP5 calculates thermal hydraulics, SCDAP calculates core heat-up, oxidation and degradation, and COUPLE is used to calculate the heat-up of reactor core material that slumps into the lower head of the RPV and is subsequently represented as debris.

- Lower head molten pool thermal analysis solver

Since it is computationally expensive to simulate this complex behavior by solving a set of Navier-Stokes equations, a number of models using lumped parameter methods and distributed parameter methods were developed to construct a computationally effective and sufficiently accurate simulation platform. PECM is used in the present study to predict the thermal behavior of the molten pool.

- Coupling method

A coupling interface, OpenMPI is used in this study due to its advantage of development cost and the calculation time. OpenMPI is a Message Passing Interface (MPI) library, which can be used for parallel calculation and coupling of solvers. In order to utilize the parallel run and information exhange through OpenMPI, several MPI functions are needed to be used in the codes for initialization, finalization of the calculation, and sending and receiving message (Fig. 10)

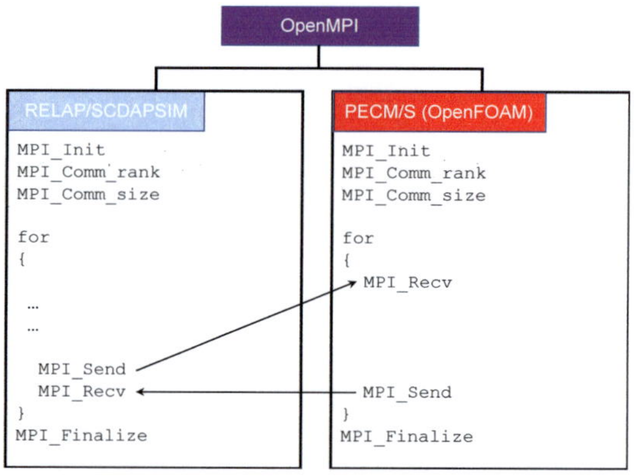

Fig. 10 Coupling of RELAP/SCDAPSIM with PECM/S.

The schematic image of the nodalization for the coupled analysis is displayed in Figure 11. The external cooling vessel is realized by pipe/annulus components (components 200 and 201) of RELAP5 and the vessel wall is assumed to be vertically connected on the side of the volumes. Heat structure components are used to represent the outer wall of the cooling vessel. A single volume (component 101), considering the amount of the molten pool, is used for the lower plenum and two volumes are attached to it for the water/air inlet and outlet, respectively. The heat transfer of molten pool and the vessel wall is calculated by PECM/S. The heat transfer coefficients at the melt surface, the external wall and the inner wall above the melt surface are calculated by the heat structure package of RELAP5.

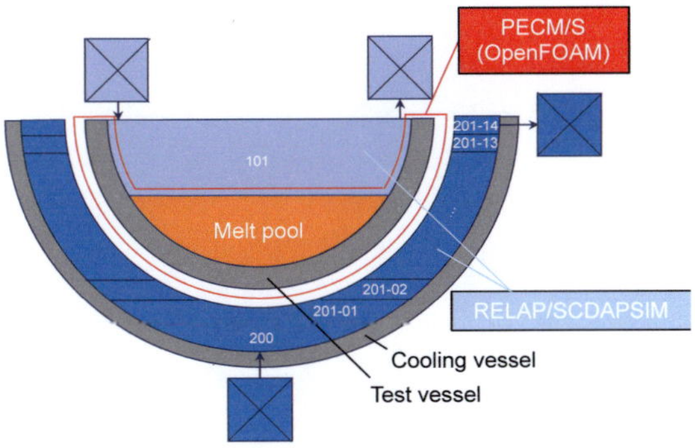

Fig. 11 Nodalization of the coupled analysis of a LIVE test

The validation has been conducted using LIVE L1, L7V and L6 tests, which have different cooling conditions and melt pool formation. The heat flux, wall temperature and crust thickness of the coupled analysis is the steady-states were compared with the experiment and with the RELAP/SCDAPSIM stand-alone calculation. A better agreement with the experimental values was obtained with the coupled analysis. For example in Figure 12, the heat flux I calculated by R/S-PECM/S was smaller than that of R/S in the lower vessel region and closer to the experimental data. The location and the value of maximum heat flux was also well predicted.

In summary, as shown by the validation analyses, the coupled approach significantly improved the prediction of the lower head thermal behavior in comparison to the RELAP/SCDAPSIM standalone calculation. Since PECM/S includes the mechanical analysis models, the mechanical behavior of the vessel can also be calculated in future applications to a real-scale severe accident analyses.

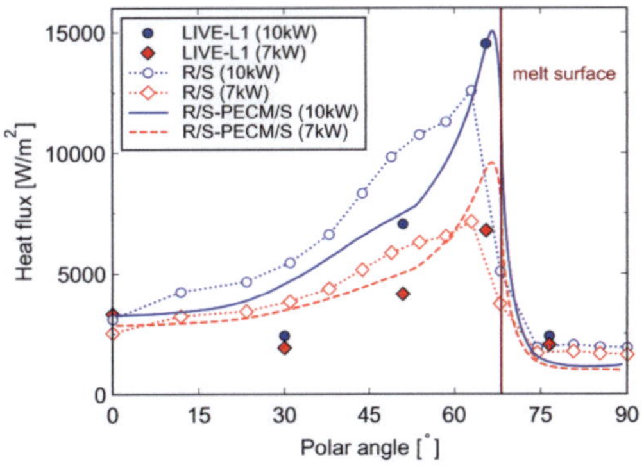

Fig. 12 LIVE-L1 heat flux profiles

References

[1] Gaus-Liu, X.; Miassoedov, A.; Influence of boundary conditions, vessel geometry, and simulant materials on the heat transfer of volumetrically heated melt in a light water reactor lower head. Journal of Nuclear Engineering and Radiation Science, Vol. 3, 031003-1, (2017).

[2] Gaus-Liu, X.; Miassoedov, A.; Journeau, C.; Liao, Y.; Cassiaut-Louis, N.; The Progress of ALISA Project: Access to Large Infrastructures for Severe Accidents in Europe and in China. ERMSAR 2017. Warsaw, Poland (2017).

[3] Gaus-Liu, X.; Miassoedov, A.; Pandazis, P.; Hollands, T.; Investigation of molten corium behavior in lower head under external subcooling and boiling conditions within LIVE experimental program and ATHLET-CD simulation. The 17th International Topic Meeting on Nuclear Reactor Thermal Hydraulics (NURETH-17), (S. 20755). Xi`an China (2017).

[4] Albrecht G.; Meyer, L.; Meignen, R.; Miassoedov, A.; Ex-vessel fuel-coolant interaction experiment in the DISCO facility, NUTHOS-12, Qingdao, China (2018)

[5] Madokoro H.; Kretzschmar F.; Miassoedov A.; "Modelling of RPV lower head under core melt severe accident condition using OpenFOAM," in 2017 International Congress on Ad- vances in Nuclear Power Plants (ICAPP 2017) – A New Paradigm in Nuclear Power Safety, Fukui-Kyoto, Japan, 2017, pp. 507–513.

[6] Madokoro H.; Miassoedov A.; Schulenberg T.; "A thermal structural analysis tool for RPV lower head behavior during severe accidents with core melt," Mechanical Engineering Letters, vol. 4, pp. 18-00038, 2018. [Online]. Available: https://doi.org/10.1299/mel.18-00038

[7] Madokoro H.; Miassoedov A.; Schulenberg T., "Assessment of a lower head molten pool analysis module using LIVE experiment," Nuclear Engineering and Design, vol. 330, pp. 51– 58, 2018. [Online]. Available: https://doi.org/10.1016/j.nucengdes.2018.01.036

[8] Madokoro H., Miassoedov A., Schulenberg T.; "Coupling of a reactor analysis code and a lower head thermal analysis solver," ASME Journal of Nuclear Engineering and Radiation Science (accepted)

[9] Miassodov, A.; et al., "The SAFEST Project: towards Pan-European Lab on Corium Behavior in Severe Accidents. Main Objectives and R&D Priorities", ERMSAR 2017, Warsaw, Poland, 2017, Paper 517.

[10] Foit, J. et al., "MOCKA-SSM experiments on basaltic concrete with and without rebars", Proceedings of NUTHOS-12, Qingdao, China, 2018

Group: Multi Phase Flow

Detailed investigations on flow boiling of water up to the critical heat flux

Giancarlo Albrecht, Stephan Gabriel

Introduction

The research group Multiphase Flows (MPF) emerges 2018 together with the group Severe Accident Research (SAR) from the Accident Analyses group (UNA) and deals with technically relevant, complex multiphase flows. During the last two years, the activities concentrated on the construction of the high-pressure test facility COSMOS-H and the experiments on the low-pressure test facility COSMOS-L, which were carried out in cooperation with the Institute for Thermal Process Engineering (TVT) at KIT and partners from industry and research.

The subject was flow boiling under forced convection up to critical heat flux (CHF). Boiling media are often used in heat exchangers to enable effective cooling of technical systems. However, when the critical heat flux is reached, the boiling crisis occurs and a sudden rise in temperature will follow, which can lead to a possible damage to system components.

Numerous experiments on COSMOS-L and preliminary tests were carried out for this purpose. Additionally, our special measurement techniques such as fibre optic void sensors or videometric void measurement technique (OVM) have also been enhanced. The experiments were carried out on the one hand in an annular test section and on the other hand in a rod bundle test section. At the same time, a test track was built up for the high-pressure tests and the construction of the high-pressure facility COSMOS-H was continued.

Experiments on Boiling under forced convection up to critical heat flux

At the COSMOS-L test facility, deionised water is used as test medium to investigate boiling crisis for various operating conditions and test section configurations. First experiments were carried out in the annular test section with one heated tube of 32 cm length made from zircalloy-4. Water is pumped through the annular gap with defined pressures and inlet temperatures while the cladding tube is electrically heated. This results in heat transfer between the cladding tube and the surrounding water up to flow boiling and critical heat flux. The system pressure was varied between 0.1 – 0.3 MPa. The maximum heat power of the test section was 17 kW; this corresponds to a heat flux of 1.78 MW/m².

A second test series was carried out in a new rod bundle test section consisting of five tubes. Mass flow and pressure were slightly reduced in these experiments, but the critical power was significantly higher. The test section assembly consists of a central tube surrounded by four additional tubes. All tubes are individually heated and instrumented. The central tube gets a moderately higher heating power, which allows that the instrumentation can be concentrated on the center tube.

Experiments in the annular test section

Extensive experiments were carried out to determine CHF statistics at the annular gap. The first experiment consisted of a CHF statistic.

Criteria for persistence and reproducibility were identified and the measurement uncertainty of the instrumentation was determined. Due to the possibility for a quick load shedding at COSMOS-L, 55 tests could be carried out in series with the same tube without any damage. The results are shown for one parameter set in the diagram in Figure 1. It shows that CHF is not a single value, but appears as a distribution.

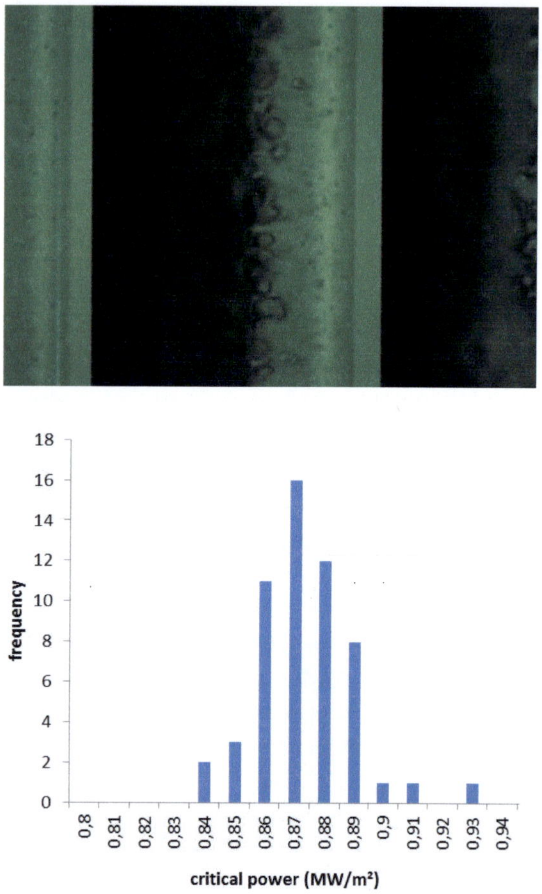

Fig. 1: Flow image and CHF-statistics over 55 single CHF Events

Further measurements were carried out with (laser-) optical fibre sensors at the annular gap to determine the local vapour content. The required optical fibre probes were manufactured, tested and validated. Measurements were performed on four parameter sets. In addition to the transformer power for heat input, the positions of the optical fibre probes were varied radially and axially, resulting in a total of 21 measuring positions. An example of the results is shown in Figure 2. The diagram shows the local vapor content as a function of the radial distance to the tube surface. The evaluation of these measurements showed that at a high power input in the upper area of the annular gap steam clusters flow along at some distance from the heating rod. As the power input decreases, these steam connections become significantly smaller. In the middle and lower area of the annular gap are no steam clusters visible. A reduced subcooling temperature hardly changes the flow conditions in the annular gap. This can be seen from the significant similarity of the content of the local steam content. However, the absolute values are lower at lower subcooling temperatures. An increase of the mass flow leads to the fact that a high density of steam bubbles occurs directly at the heating rod only in a higher position of the annular gap. With the exception of the local steam content in the middle area near the heating rod, however, the values remain in comparable ranges with a mass flow increase. If the system pressure is reduced, the steam clusters mentioned already occur with a slightly lower power input.

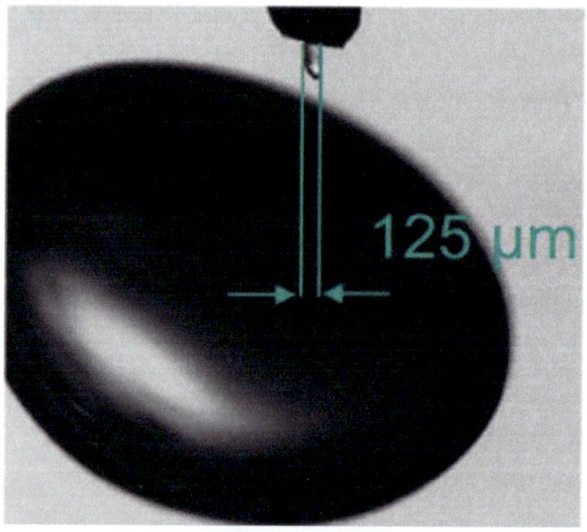

Fig. 2: Image of bubble and probe tip (right) phase distribution near the heater surface.

Furthermore, the decrease of the local steam content with increasing distance to the heating rod is lower from such a power input, which leads to the conclusion that the produced steam bubbles collapse later at lower system pressure and can move further into the radius of the annular gap.

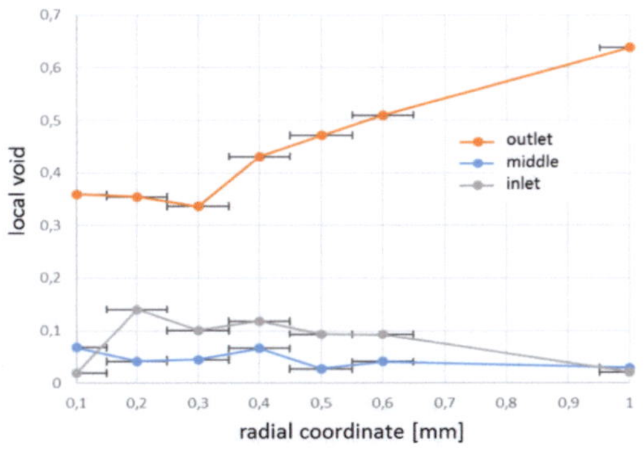

Fig. 3: Image of bubble and probe tip (right) phase distribution near the heater surface

Another important part of the investigations at the annular gap were the laser-based measuring methods PIV and shadowgraphy at the annular gap test section. Measured values were bubble velocity (axially and radially), bubble velocity fluctuation, bubble position and bubble size. An excerpt from the results can be seen in Figure 3.

The velocity profile shows a significant acceleration of the flow from bottom to top through the decreasing fluid density with increasing steam mass fraction. In addition, the measuring positions at the inlet and in the center of the test section are showing a significantly larger boundary layer on the heated tube wall. The bubble size distribution shows the spectrum of gas bubbles in the flow. Both vapor bubbles and bubbles of non-condensable gases are detected if they still exist. In addition to the size distribution, which is an important result for the validation of CFD codes, bubble trajectories, the depth at which the bubbles penetrate into the subcooled flow and further spatially resolved data could be recorded and evaluated in these experiments. The results provided a detailed impression of the processes taking place and permitted our Project partners to validate their advanced CFD-Codes.

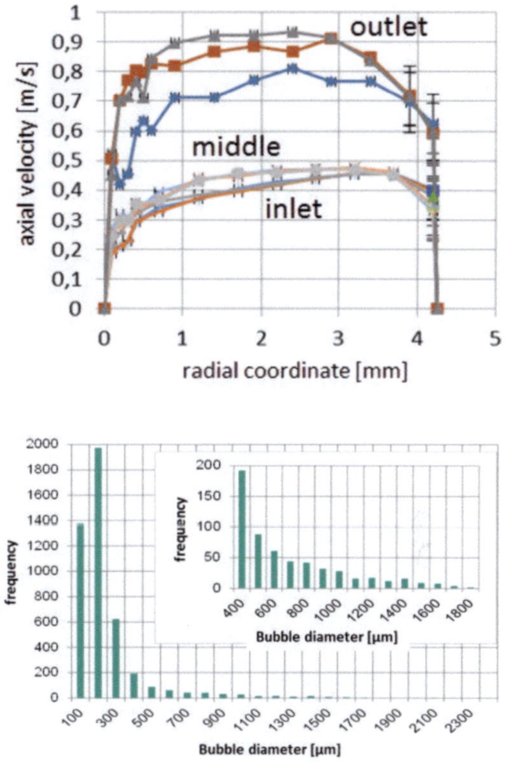

Fig. 4: Velocity distribution and droplet size distribution in the boiling flow

Experiments in the rod bundle test section

The tests on the rod bundle were carried out in an additional test section. Heating rod arrangement, spacing and material were chosen in such a way that they come as close as possible to those in a reactor. Extensive data packages were also created during these tests and made available to the project partners. The experience gained within the project concerning instrumentation, relevant parameter ranges, etc. and the completed high-pressure test section can also be applied in subsequent projects. Figure 4 gives an Impression of different flow conditions. It becomes visible how strongly the two-phase flow mixes in the test track. The flow pulsates violently at high heating powers, so that no single phase interfaces are more recognizable.

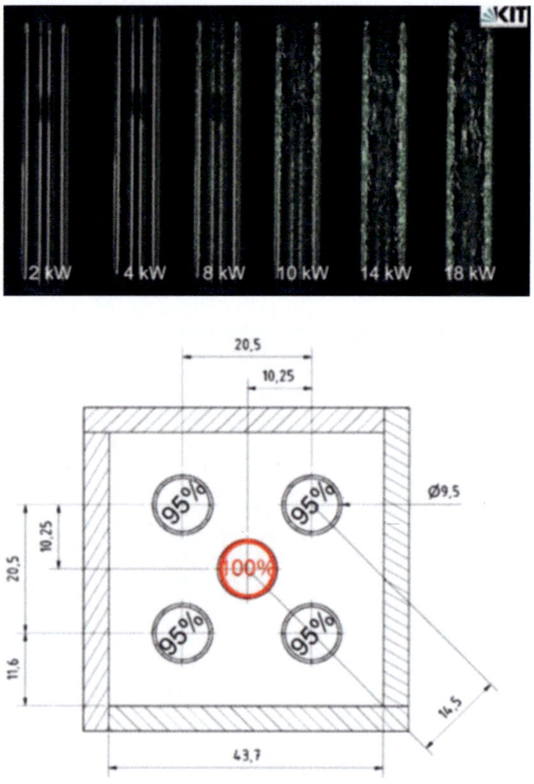

Fig. 5: (top) flow image of a rod bundle image (bottom) cross section of the rod bundle test section

Due to the high grade of instrumentation, detailed data could nevertheless be collected in these flows. An example of this is the temperature distribution of the heated tube wall, which can be seen in Figure 5. Measurements with the above-mentioned fiber probe and laser doppler anemometry (LDA) are currently being carried out.

High Pressure-Loop

The COSMOS-H plant was developed for the investigation of safety-relevant thermohydraulic phenomena under reactor-typical conditions. Both the modular test track (pressure hull) and the required assembly trolley to precisely arrange, assemble and disassemble the test section were completed. The planned test section for high-pressure tests, including the optical access modules, was designed and manufactured. The test section provides all the required characteristics and passed the pressure test according to the Pressure Equipment Directive at 32 Mpa (Fig. 7). The crane traverse required to lift the 1.6 tonne test section without bending and to turn it into the installation position is constructed and is currently in production.

The construction of the high-pressure loop continues, the two cooling loops have been completed and passed the pressure test. The power supply for the test track has already

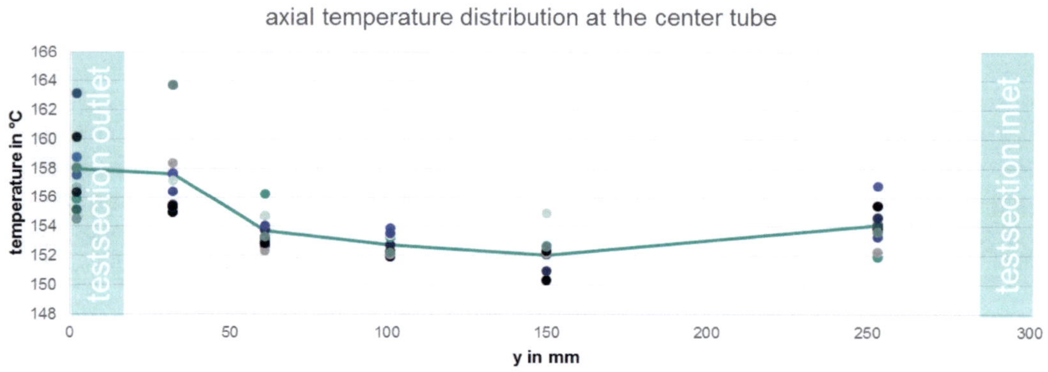

Fig. 6: Axial temperature distribution of the central tube in the rod bundle

been put into operation and has passed the load-loss tests carried out.

Fig. 7: (top) Ground floor at construction Site COSMOS-H, (bottom) modular high pressure test section for COSMOS-H

Conclusions and Outlook

Overall, good progress has been made over the last two years in understanding flow boiling and boiling crises. Experimental facilities and Measurement techniques have been further developed and are now incorporated in new projects. Nevertheless, much work remains to be done.

Recent work is the further adaptation and further development of the measuring technology in relation to the glass fiber probes, optical void measurement as well as the laser measuring technique for disperse and continuous flow components. Furthermore, a method for locating CHF events by triangulation of thermocouple signals is currently being developed.

The construction of the high-pressure facility COSMOS-H will be continued with the completion of the high-pressure circuit within the next year. Furthermore, a new project for the critical heat flux in intermittent flows started in January 2019 at COSMOS-L.

Partners

We thank Mr. Florian Kaiser, Institute for Thermal Process Engineering (TVT), KIT Karlsruhe and Mr. Wilson Heiler, ProScience, Ettlingen, for their dedicated work on the experimental facilities and in the project NUBEKS.

List of Acronyms

MPF	Group Multiphase Flows
SAR	Group Severe Accident Research
UNA	Group Accident Analysis
TVT	Institute for Thermal Process Engineering
CHF	Critical Heat Flux
OVM	Optical Void Measurement
WENKA	Water Entrainment Channel Karlsruhe
COSMOS-L	Critical heat flux On Smooth and Modified Surfaces - Low pressure loop
COSMOS-H	Critical heat flux On Smooth and Modified Surfaces - high pressure loop
LDA	Laser Doppler anemometry
PIV	Particle Image velocimetry

References

[1] Kaiser, F.; Gabriel, S.; Albrecht, G.; (2017) "Investigation of the Critical Heat Flux in an Annular Gap and a Rod Bundle Configuration under low Pressure Conditions". ProcessNet-Fachgruppe Wärme- und Stoffübertragung, Bruchsal, 16.-17. Februar 2017 (Poster)

[2] Gabriel, S.; Albrecht, G.; Kaiser, F.; (2017) "Experimental Investigations of the Critical Heat Flux in a modular Test Section under high Pressure Conditions". ProcessNet-Fachgruppe Wärme- und Stoffübertragung, Bruchsal, 16.-17. Februar 2017 (Poster)

[3] Kaiser, F.; Gabriel, S.; Albrecht, G.; Huckle, J.; (2016) "Investigation of the Critical Heat Flux in a Rod Bundle Configuration under low Pressure Conditions". Jahrestagung KIT-Zentrum Energie, Leopoldshafen, 15. Juni 2016 (Poster)

[4] Gabriel, S.; Kaiser, F.; Schulenberg, T.; Albrecht, G.; Heiler, W.; Wetzel, Th.; Miassoedov, A.; "Optical void measurement method for stratified wavy two-phase flows". 9th World Conference on Experimental Heat Transfer, Fluid Mechanics and Thermodynamics, 12-15 June, 2017, Iguazu Falls, Brazil (Presentation)

[5] Gabriel, S.; Kaiser, F.; Schulenberg, T.; Albrecht, G.; Heiler, W.; Wetzel, Th.; Miassoedov, A.; "Optical Void Measurement Method for stratified wavy two-phase Flows". Experimental Thermal and Fluid Science 97 (2018) 341–350

[6] Haas, C.; Kaiser, F.; Schulenberg, T.; Wetzel, Th.: "Critical heat flux for flow boiling of water on micro-structured zircaloy tube surfaces". International Journal of Heat and Mass Transfer, 120 (2018) 793-806

[7] Frank, Th.; Amine Ben Hadj Ali; Lifante, Conxita; Bruder, Moritz; Kaiser, Florian; Eickenbusch, Henning; "Prediction of Convective Boiling up to Critical Heat Flux (CHF) Conditions for Test Facilities with Vertical Heaters". 15th Joint HZDR & ANSYS Conference "Multiphase Flows: Simulation, Experiment and Application". Helmholtz Zentrum Dresden-Rossendorf (HZDR), Germany, 14.-17. Nov. 2017.

[8] Frank, Th.; Amine Ben Hadj Ali; Lifante, Conxita; Kaiser, Florian; Gabriel, Stephan; Eickenbusch, Henning; "Modeling of Wall-boiling Phenomena from Nucleate Subcooled Boiling up to CHF Conditions" .13. ERCOFTAC Technologietag, ERCOFTAC Pilotzentren Deutschland Süd und Nord, Internationales Begegnungszentrum "Eulenhof" der Universität Stuttgart, Germany, 30. November 2017. ERCOFTAC = European Research Community on Flow, Turbulence and Combustion.

[9] Albrecht, G.; Meyer, L.; Miassoedov, A.; Kirstahler, M.; Schwall, M.; Wachter, E.; Gaus-Liu, X.; FCI Experiments in the DISCO-H Facility performed under the COOPERATION AGREEMENT N° 45001082 between IRSN and KIT, (2015)

[10] Höhne, T.; Gabriel, S.; "Simulation of a counter-current horizontal gas/liquid flow experiment at the WENKA channel using a droplet entrainment model", Annals of Nuclear Energy 121, 414-425, (2018)

[11] Albrecht, G.; Meyer, L.; Meignen, R.; Miassoedov, A.; Ex-vessel fuel-coolant interaction experiment in the DISCO facility in the LACOMECO project, J. Nuclear Science and Technology, Volume 53, (2016)

[12] Albrecht, G.; Meyer, L.; Meignen, R.; Miassoedov, A.; Ex-vessel fuel-coolant interaction experiment in the DISCO facility, NUGENIA TA2 Meeting, Spain (2016)

[13] Albrecht, G.; Meyer, L.; Meignen, R.; Miassoedov, A.; Ex-vessel fuel-coolant interaction experiment in the DISCO facility, NUTHOS-12, Qingdao, China (2018)

[14] Albrecht, G.; Meyer, L.; Meignen, R.; Miassoedov, A.; Ex-vessel fuel-coolant interaction experiment in the DISCO facility, SAFEST final seminar, Cadarache, France (2018)

Group: Karlsruhe Liquid metal LAboratory

Carbon dioxide free production of hydrogen

The following paper is a reproduction of the contribution **"Carbon dioxide free production of hydrogen" by L Stoppel, T Fehling, T Geißler, E Baake and T Wetzel, published in** *IOP Conf. Ser.: Mater. Sci. Eng.* **228** 012016, as accessible under DOI 10.1088/1757-899x/228/1/012016, using the Creative Commons Attribution 3.0 license.

Abstract

The present report summarizes the theoretical modelling and experimental investigation results of the study on the direct thermal methane cracking. This work is a part of the LIMTECH-Project (Liquid Metal Technologies) funded of Helmholtz Alliance and was carried out from 2012 to 2017. The Project-part B5 "CO_2-free production of hydrogen" focused on experimental testing and particularly on modelling the novel methane cracking method based on liquid metal technology. The new method uses a bubble column reactor, filled with liquid metal, where both the chemical reaction of methane decomposition and the separation of gas fraction from solid carbon occur. Such reactor system was designed and built in the liquid metal laboratory (KALLA) at KIT. The influences of liquid metal temperature distribution in reactor and feed gas flow rate on methane conversion ratio were investigated experimentally at the temperature range from 930°C to 1175 °C and methane flow rate at the reactor inlet from 50 to 200 mLn/min. In parallel with experimental investigations, a thermochemical model, giving insight in the influence of the above mentioned parameters has been developed at KIT and a CFD model was developed at LUH to get an overview about the bubble dynamics in the reaction system. The influence of different bubble sizes and shapes, multi-inlet coalescence effects as well as the potential of electromagnetic stirring have been investigated.

Introduction

The carbon dioxide emission (CO_2) from human activities influences the earth's ecosystem significantly. The worldwide economic growth and development plays a key role in the upward trend in CO_2 emission. As a result, there are many ongoing research works on development of efficient technologies that will be able to reduce the CO_2 emission. One of the technologies, whose usage led to producing a considerable amount of carbon dioxide, is the steam methane reforming (SMR). Depending on the feedstock, this technology provides production of 9 to 12 tons of CO_2 as a by-product per one ton of produced hydrogen. In the future, current SMR process might be replaced by other environment-friendly hydrogen production technologies based on carbon free sources. However, from the economic point of view, it is not to expect that in the near future hydrogen will be completely produced from renewable and sustainable resources. The proposed direct methane cracking method can in relatively short time provide an attractive industrial solution that leads to the reducing of global carbon dioxide emission while using fossil feedstock, i.e. natural gas.

The chemical reaction of the direct methane cracking can be described by an overall endothermic reaction:

$$CH_4(g) \rightarrow C(s) + 2H_2(g) \quad \Delta H_R^0 = +74.85\frac{kJ}{mol} \tag{1}$$

The reaction products are gaseous hydrogen and solid elemental carbon. The energy requirement for this chemical reaction is moderate, but a considerable shift of the thermodynamic equilibrium in direction of the reaction products is only achievable beyond 800°C (1 atm). At 1200°C, the theoretical methane conversion efficiency is about 95%.

The direct thermal decomposition of methane is not a new idea, see [2-4] for further reference. In the past, a lot of research works were dedicated to the investigation and realizing this process experimentally. The main problem in the practical implementation was the very short operating time of the developed systems. Indeed, the produced carbon is deposited in the reaction zone and this causes the blockage of the reactor. In comparison to previous approaches, the main advantage of the suggested method is the continuous separation and transport of the produced solid carbon from the reaction zone. It occurs owing to a difference of density between liquid metal and carbon and it is supported by the rising gas bubbles. After reaching the liquid metal surface, the carbon powder can be removed from the system with available conventional methods used in the industry (skimming, gas floatation, etc.).

Due to the opaqueness of the liquid metal, visualization of bubble dynamics in the reactor is hardly possible. Numerical simulations have therefore been done in order to investigate dependencies of bubble residence times, bubble shapes, potential carbon sediment locations etc. on operation parameters like gas flow rate. To identify potential optimizations, reactor geometry, inlet types and the heating system have been varied. Especially in this article the aspect of multi-inlets as well as the effect of electromagnetic stirring with regards to increase of hydrogen yield is put in the focus.

Basic process description and experimental set-up

The simplified process flow diagram is shown in Figure 1. The methane cracking reaction occurs in a vertical bubble column reactor filled with liquid metal. Natural gas or pure methane can be used as the feed gas. It is injected at the bottom of the column reactor. The gas bubbles formed at the injector rise to the top of the reactor. Thereby, methane is heated up to liquid metal temperature level and split into gaseous hydrogen and solid carbon.

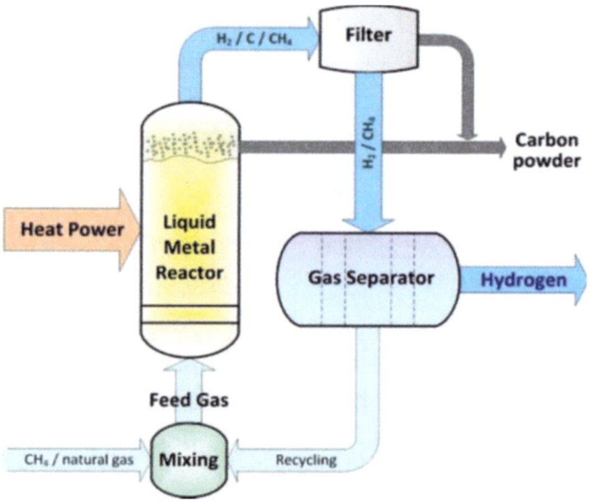

Fig. 1. Process flow diagram [1].

The separation of produced hydrogen from the methane-hydrogen mixture takes place in a separator. The unreacted methane and other possible gaseous by-products can be mixed with the feed gas via mixing unit and moved again to the reactor inlet. Most of the solid carbon formed in the cracking reaction collects on the liquid metal free surface and can be removed from the reactor relatively easy. The filter installed within the outlet gas line holds back residual, gas-borne carbon particles.

The construction of experimental test facility HELiS (Hydrogen production Experiment in Liquid Sn) has been part of this project and has been operated at the KArlsruhe Liquid metal LAboratory (KALLA) of KIT. In particular, the facility consists of a gas supply system, an experimental port for the liquid metal reactor, an outlet gas analysis system and an instrumentation and control system. The gas supply system is designed for feeding the working gas in required mixing proportion (methane/argon or nitrogen) into the chemical reactor. It consists of a manifold of standard cylinders with commercial high purity argon at pressure of 200 atm, pipelines, pressure gauges, and gas flow sensors. The use of liquid tin (Sn) as the working fluid in the bubble column reactor has been based upon a careful selection process along several criteria like inert behaviour towards the reaction gases. Major advantages are of course the high thermal conductivity, high density compared to carbon, non-toxicity and long-term chemical stability. The strong corrosion attack of liquid tin at high temperatures towards metallic, particularly steel container materials poses special challenges, which will be detailed in the following.

During the study, two types of the reactor design have been developed and tested: stainless steel reactor and combined quartz glass-steel reactor designs. In general, stainless steel is the favoured construction material due to the great design flexibility and the availability of connectors and flanges even for high temperature applications. However, in contact with molten tin, stainless steel alloying elements are dissolved increasingly with increasing temperature. In the experimental setup, 316 and 1.4549 steels were used as reactor materials. In both cases, experiments showed that materials are corroded quite fast. The lifetime of these reactors was insufficient. In comparison with stainless steel, the quartz glass is chemically stable and showed no corrosion impact towards tin within the planned temperature range. The main idea of this reactor design concept is to couple the advantages of the chemically stable glass material and the mechanical stability of stainless steel. The non-

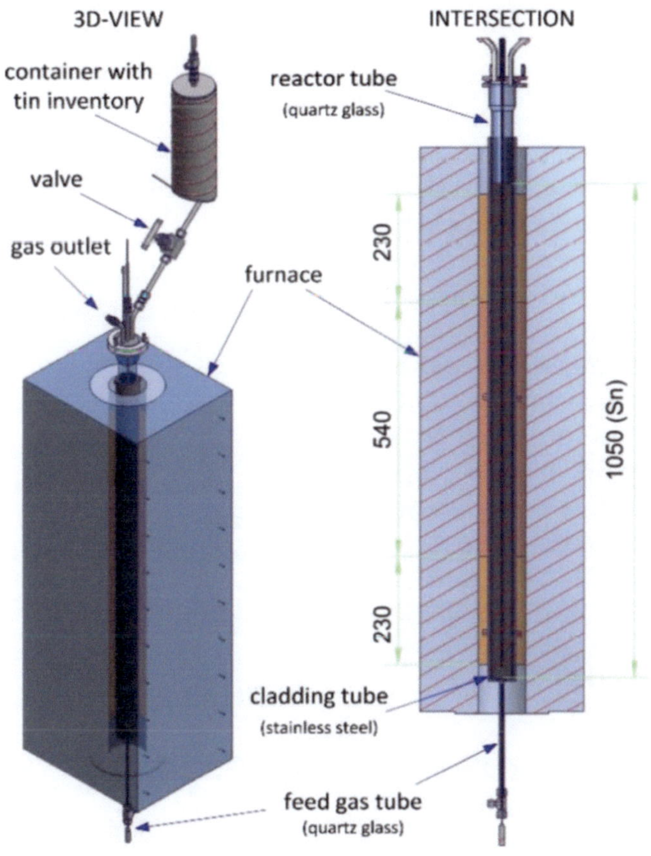

Fig. 2. Experimental set-up.

corrosive quartz glass tube getting in contact with liquid metal is inserted into the steel tube. This design proved reliable during the experimental campaigns [2, 3, 4, 5].

In detail, the chemical reactor used in the experiments is a vertical quartz glass tube with inner diameter of 40.6 mm and length of 1268 mm, which is inserted into a stainless steel tube with inner diameter of 49.25 mm and length of 1150 mm (Fig. 2). A single orifice (Ø 0.5 mm), at the bottom of the quartz glass tube was used as gas injector. The combined reactor was placed inside of an 8 kW electric furnace with a maximum operating temperature of 1200 °C. The upper part of the reactor, so called gas zone, is ending above the furnace and was insulated separately. The temperature inside of the reactor was continuously measured during the operating time by applying an alumina tube with type K thermocouples at 10 positions along the whole reactor length. The temperature measured at a level of 600 mm in the middle of the reaction zone was used as reference temperature. Glass fragments or cylindrical rings made of quartz glass have been used to form packed beds. The space porosity of the packed beds was varied from 76% to 84%. The separate supply tank placed above the rector was used for melting of tin prior to filling up the reactor. The gaseous reaction products were filtered via sinter metal elements made of stainless steel with pore size 0.5 µm. The analysis of product gas was realized by using a gas chromatograph (GC). During the experiments, the mole-fraction was continuously measured for methane, hydrogen, nitrogen, ethane, ethylene and acetylene at various operating conditions.

Thermo-chemical modelling and experimental results

The following section describes the thermo-chemical modelling and the experimental results and extends our previous work [3].

Thermo-chemical modelling

To estimate the influence of different parameters on the process as well as for future process description and hydrogen yield predictions, a thermo-chemical (TC) model was developed. The model is based on three coupled one dimensional partial differential equations with spherical coordinates, solved using MATLAB® parabolic PDE solver. The energy equation (3), the species equation (4) and a pressure equation (5) are solved inside of a bubble traveling through the liquid metal reactor. The bubble is assumed as a spherical body with a constant radius and a constant residence time in the reactor depending on the inventory material and height. As the bubble radius is kept constant, volume expansion due to the pressure drop and the chemical reaction along the liquid metal reactor length is not included in the model. The initial bubble radius in equation (2) is calculated by Tate´s law, which bases on a force balance between buoyancy and surface tension [6] using temperature dependent surface tension and density from Alchagirov et al. [7], [8].

$$r_{B,0} = \left(\frac{3 d_{orf} \sigma_0}{2 \rho_0 g} \right)^{1/3} \quad (2)$$

where d_{orf} is the diameter of the orifice, σ_o and ρ_o are the surface tension and density respectively.

In the thermo-chemical model, all gas properties are functions of the present local temperature, pressure and gas composition. The source terms in equation (3) and (4) imply temperature dependent reaction rates, following Arrhenius law and assuming a first order chemical reaction.

Energy equation

$$\rho_G c_{p,G} \frac{\partial T}{\partial t} = \frac{1}{r^2} \frac{\partial}{\partial r}\left(r^2 \lambda_G \frac{\partial T}{\partial r} \right) - \frac{k_0 \Delta H_R p\, y_i}{R_g T} \exp\left(-\frac{E_a}{R_g T} \right) \quad (3)$$

where ρ_G is the gas density, $c_{p,G}$ is the specific heat capacity of the gas, T is the temperature,

r is the bubble radius, λ_G is the heat conductivity of the gas, k_0 is the pre-exponential factor, ΔH_R is the reaction enthalpy, p is the absolute pressure, R_g is the universal gas constant, y_i is the mole fraction, E_a is the activation energy.

Species equation

$$\frac{p}{T}\frac{\partial y_i}{\partial t} = \frac{1}{r^2}\frac{\partial}{\partial r}\left(r^2 \frac{pD_{i,j}}{T}\frac{\partial y_i}{\partial r}\right) - \frac{k_0 y_i p}{T}\exp\left(-\frac{E_a}{R_g T}\right), \tag{4}$$

where D is the diffusion coefficient.

Pressure equation

$$\frac{\mathrm{d}p}{\mathrm{d}t} = -\frac{(p_0 - p_1)}{\tau_G}, \tag{5}$$

where τ_G is the residence time.

Assuming linear pressure drop from the bottom to the top of the reactor, whereby the local liquid metal hydrostatic pressure in the column equals the pressure inside of the bubble. The hydrostatic pressure is calculated using the temperature dependent tin density equation from Alchagirov et al. [8].

The feed gas conditions at the bottom of the reactor are used as the initial conditions in equation (3) – (5).

$$T = T_0; \qquad y_i = 0; \qquad p = p_0 \tag{6}$$

Symmetry boundary conditions are set at the center of the bubble for the bubble temperature, the species fraction and the pressure (equation (10)). At the bubble interface, symmetry boundary conditions are set for the species fraction and the pressure (equation (11)). As the bubble is rising in the liquid metal due to density differences, and assuming the reference coordinate system fixed to the travelling bubble, the liquid metal is flowing around the bubble. Assuming further, that gas convection inside of the bubble can be neglected and the bubble behaves like a solid sphere, the following Nusselt correlation for forced convection over a sphere from Melissari et. al [9] can be used to model heat transfer from the tin to the bubble. The correlation is valid in a wide range of Pr and Re numbers, including the present Pr number for tin (Pr = 0.0031) at the lowest investigated temperature.

$$Nu = 2 + 0.47 Re^{0.5} Pr^{0.36}, \tag{7}$$

with

$$Re = \frac{u_B d_B}{\nu_{LM}}, \tag{8}$$

where u_B is the bubble velocity, d_B is the bubble diameter, ν_{LM} is the kinematic viscosity of the liquid metal.

Even at low Re numbers, resulting from small bubble rise velocities and diameters, the Nu number always leads to a value of 2 or higher. Evaluating the heat transfer limitation, the Bi number in equation (9) reveals a minimum value of 520 for the highest investigated temperature

$$Bi = \frac{Nu \lambda_{Sn}}{\lambda_{CH_4}}, \tag{9}$$

where λ_{Sn} is the heat conductivity of tin and λ_{CH_4} is the heat conductivity of methane.

Consequently the heat transfer limitation is on the gas phase side and the interfacial bubble temperature is set to the local liquid metal temperature and changes with time, and thus with the liquid metal height.

Inside of the bubble ($at\ r = 0$), the conditions are defined as follows

$$\frac{\partial T}{\partial r} = 0; \qquad \frac{\partial y_i}{\partial r} = 0; \qquad \frac{\mathrm{d}p}{\mathrm{d}r} = 0 \tag{10}$$

On the gas liquid interface of the bubble ($at\ r = r_B$), the conditions are defined as follows

$$T = T_{LM}(t); \quad \frac{\partial y_i}{\partial r} = 0 \quad ; \quad \frac{\mathrm{d}p}{\mathrm{d}r} = 0 \tag{11}$$

To estimate the performance of the process using the measured gas mole fractions from the experiments, the hydrogen yield, the methane conversion and the ratio between the produced hydrogen and the reacted methane are investigated. As the product gas analysis in the GC only considers the components in the gas phase, without taking into account the produced solid carbon, basic calculations to clarify the difference between the measured mole fractions in the GC and the mole fractions of the components, considering carbon production, are necessary. For pure methane as feed gas, assuming the behaviour of an ideal gas, the initial total molar flow rate is given by

$$\dot{N}_{total,0} = \dot{N}_{CH_4,0} = \frac{p_0 \dot{V}_{CH_4,0}}{R_g T_0}, \tag{13}$$

the molar flow rates of the different product gas components, considering no intermediates or by-products, are as follows

$$\dot{N}_{H_2,1} = 2X_{CH_4,1}\dot{N}_{total,0} \tag{14}$$

$$\dot{N}_{CH_4,1} = \dot{N}_{total,0}\left(1 - X_{CH_4,1}\right) \tag{15}$$

$$\dot{N}_{C,1} = X_{CH_4,1}\dot{N}_{total,0} \tag{16}$$

Using Equations (14) - (16), the total molar flow rate at the outlet of the reactor is calculated by

$$\dot{N}_{total,1} = \dot{N}_{total,0}\left(1 + 2X_{CH_4,1}\right) \tag{17}$$

To estimate the output mole fraction of each component the following term is valid

$$y_{i,1} = \frac{\dot{N}_{i,1}}{\dot{N}_{total,1}} \tag{18}$$

Usually, given by mass conservation, the sum of all components present in the chemical reaction leave the reactor. In this process, the carbon stays inside of reactor, which reduces the total mass and consequently the total molar flow of the product gas. This leads to a new total molar flow entering the GC, changing the mole fraction of each component in the product gas, without taking the carbon mole fraction into account. With the new GC molar flow rate, the calculation of the methane conversion, using GC mole fraction data and Equations (14) - (18), result in

$$X_{CH_4} = \frac{1 - y_{CH_4,GC}}{1 + y_{CH_4,GC}}, \tag{19}$$

whereby the hydrogen yield is given by

$$Y_{H_2,CH_4} = \frac{\dot{N}_{H_2,1}}{2\dot{N}_{CH_4,0}} = \frac{y_{H_2,GC}\left(1 + X_{CH_4}\right)}{2} \tag{20}$$

Besides the hydrogen yield, the ratio between the produced hydrogen and the reacted methane is mandatory to estimate possible deviations due to the formation of intermediates

$$\frac{\dot{N}_{H_2,prod}}{2\dot{N}_{CH_4,reac}} = \frac{y_{H_2,GC}\left(1 + X_{CH_4}\right)}{1 - y_{CH_4,GC}\left(1 + X_{CH_4}\right)} \tag{21}$$

This ratio leads to a value of 2, in case of all the methane in the feed gas converts to hydrogen and carbon without the formation of intermediates.

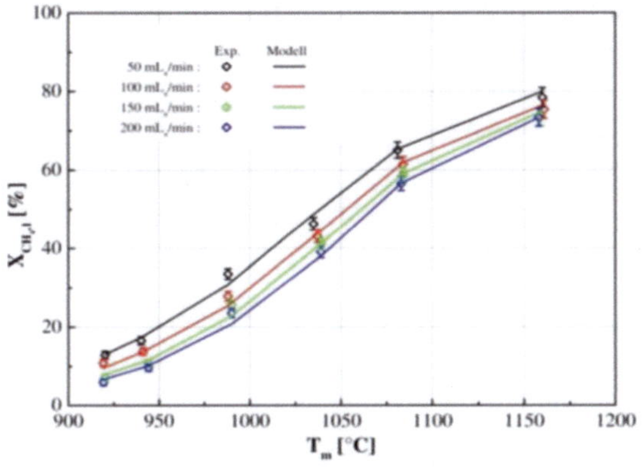

Fig. 3. Measured and calculated methane conversion as a function of the average temperature [2].

Comparison of calculation and measurements

As mentioned above, to quantify the chemical process of the direct methane cracking in liquid metal, several experimental campaigns were performed at the liquid metal laboratory (KALLA). The hydrogen yield measured by the gas chromatograph was used as an indicative parameter to estimate the potential of the process. At the experiments, special focus was given on the influence of the feed gas volume flow rate at different liquid metal temperatures and on the reactor clogging due to carbon formation.

The figure 3 shows the experimental results of methane conversion compared to calculated data as a function of the average liquid metal temperature in reactor at different feed methane volume flow rates. The model prediction showed very good agreement with the experiments.

Especially the liquid metal temperature has a significant influence on the resulting methane conversion, whereby the dependency of the applied methane volume flow rate is quite moderate. The influence of the gas residence time in the liquid metal reactor is only slightly visible by changing the methane feed volume flow rate as the packed bed itself enhances the residence time significantly compared to a reactor design without a packed bed. Nevertheless, applying a pure methane volume flow rate of 200 mLn/min at 1175 °C results in an almost 4 times higher overall hydrogen standard volume flow rate at the outlet compared to a hydrogen output applying a 50 mL_n/min initial methane feed volume flow rate.During the experiments, almost no intermediate reaction products occurred, apart from a maximum mole fraction of 0.5 mol-% ethane and 1.0 mol-% ethylene in the product gas could be detected, whereby no acetylene was measured. As a consequence, the hydrogen selectivity in this process is almost one, leading to a potential product portfolio of only hydrogen and solid carbon.

While disassembling the quartz glass reactor, a thin carbon layer with a thickness in the range of 10 µm was found between the inventory and the quartz glass reactor wall. In the upper part of the reactor, above the metal surface, a mixture of the packed bed and the produced carbon, appearing as a powder, was found (Figure 4). By means of the observed amount, most of the produced carbon accumulated in this area. A smaller amount, transported by the product gas, was collected in the filter elements between the off-gas tube and the GC.

Fig. 4. Carbon powder produced during the experiment [1].

Numerical investigation of fluid dynamic effects within the reactor

To get an idea about the bubble dynamics inside the reactor system of the HELiS facility and to evaluate potential optimization opportunities, additionally to the experiments numerical simulations have been done. According to an efficient production, the dynamics and residence times for different bubble diameters as well as optimization methods like electromagnetic stirring or multi-inlet usages were discovered. Simulations were done using the open source software package *OpenFOAM*.

Phase description models for numerical bubble simulation

To describe the physics of a two phase bubble flow and to get a numerical description of the dynamics, generally, two different methods are available. The discrete phase models for particle tracking based on *Euler-Lagrange* description and *Euler-Euler* approaches with a Eulerian description based on an inertial observer formulation for both phases. The Euler-Euler-approaches can be also subdivided into Volume of Fluid, mixture and Euler methods. Prior simulations, especially magnetohydrodynamic modeling have been done using Volume of Fluid method. In the Volume of Fluid method, a volume fraction equation is solved to calculate volume-fraction-averaged parameter that is used to solve the momentum equation. A detailed explanation can be found in [10].

In actual simulations the Eulerian model, described in [11], is used. A solver for coupling this method also with electromagnetic calculations in the software *getDP* has been developed. One advantage is a more exact calculation for greater bubble volume fractions and greater Stokes numbers, especially at the surface and for different velocities $\vec{v}$ of the phases. In contrast to the Volume of Fluid model the Eulerian model solves a separated set of equations for every phase. The coupling between the phases is reached by pressure- and phase-exchange coefficients. The continuity equation, given by

$$\frac{\partial \alpha_q}{\partial t} + \vec{v} \cdot \nabla \alpha_q = \frac{1}{\rho_q}\left(\sum_{p=1}^{n} \dot{m}_{pq} - \dot{m}_{qp}\right) + S_q \tag{22}$$

is used to solve the volume fraction α_q in the momentum equation for every fluid phase:

$$\frac{\partial}{\partial t}\left(\alpha_q \rho_q \vec{v}_q\right) + \nabla \cdot \left(\alpha_q \rho_q \vec{v}_q \vec{v}_q\right) = -\alpha_q \nabla p + \nabla \cdot \bar{\bar{\tau}}_q + \alpha_q \rho_q \vec{g} + \alpha_q \rho_q \left(\vec{F}_q + \vec{F}_{lift,q} + \vec{F}_{vm,q}\right) + \sum_{p=1}^{n}\left(K_{pq}\left(\vec{v}_p - \vec{v}_q\right) + \dot{m}_{pq}\vec{v}_{pq}\right) \tag{23}$$

Here ρ_q describes the density, ∇p the pressure gradient and $\dot{m}_{pq}$ the mass transfer between phase p and q. The qth phase stress tensor $\bar{\bar{\tau}}_q$ is a function of phase fraction and velocity. External forces $\vec{F}_q$, lift forces $\vec{F}_{lift,q}$ and virtual mass forces $\vec{F}_{vm,q}$ can be calculated by correlations described in [12]. The phase exchange coefficient K_{pq} between the dispersed phase q (methane) and continuous phase p (tin) is a function of drag and will be calculated by a *Schiller-Naumann-Correlation* [13] in the following studies. For the description of energy transfer between the two phases, a separate enthalpy equation has to be solved for each phase:

$$\frac{\partial}{\partial t}\left(\alpha_q \rho_q h_q\right) + \nabla \cdot \left(\alpha_q \rho_q \vec{v}_q h_q\right) = \alpha_q \frac{\partial p_q}{\partial t} + \bar{\bar{\tau}}_q \div \nabla \vec{v}_q - \nabla \cdot \vec{q}_q + S_q + \sum_{p=1}^{n}\left(Q_{pq} + \dot{m}_{pq} h_{pq}\right) \tag{24}$$

The parameter h_q describes the enthalpy, S_q includes possible source terms. In addition, it includes the heat flux Q_{pq} that is a function of heat transfer coefficient between the phases. It is dependent on the Nusselt number. In the following studies the correlation from *Ranz and Marshall* [14] is used for its description. The fluid dynamics in the system methane and tin are turbulent. To describe the turbulent flow structures in the system a *mixture-k-epsilon-model* from Behzadi et.al [15] is used. The main idea of this model is based on the assumption that turbulence is dictated by the continuous phase, so turbulent kinetic energy and dissipation for each phase can be linked with constants.

Bubble size studies and slug flow dynamics

In the following studies simulation results of bubble flow in reactor using an Eulerian model will be described. In this section bubble residence times will be analyzed as a function of bubble diameters. As well as the system temperature the residence time is also coupled with the hydrogen yield and offers a potential for optimization. In these models, a tin temperature of 900°C is assumed. Varying the temperature would also influence the rates of ascents. Generally, it has been shown that carbon sediments occur also at walls near the bottom of the reactor, so a greater range from 1 mm bubbles up to bubble sizes near the reactor diameter were discovered (see figure 5.b-d.). The maximum possible bubble size due to reactor diameter is limited to 40 mm. The results of the diameter studies are shownin figure 6. According to experimental results with the system air/water, a spherical bubble shape occurs at smaller bubble diameters, which results in lower rates of ascent and respectively higher residence times. For the spherical shape, rates of ascents increase with bubble diameter until the point of 7 mm. For greater diameters, oscillations occur that results in lower bubble velocities. Further increasing the bubble diameter leads to bubble shape change due to greater Eötvös-number that increases the velocity again. For bubble diameters greater than 20 mm another effect develops. As it can be seen in figure 5.b, stronger wall effects occur, especially at the lower part of the reactor, these effects also lead to bubble dispersion. This is also a possible explanation for carbon sediments occurrence at the reactor walls near the lower part of the reactor and the described velocity reduction. After dispersion effects at the lower part of the reactor, coalescence occur after a short time, so a single bubble leaves the melt after the described residence time (figure 5.c).

For the simulation of constant volume flows used in experiment, the effect of generating greater bubbles and splitting can be seen, as well as the occurrence of wobbling with increasing volume flow, which also fits to other experiments. Another opportunity of comparison, the change of inclination angle of the reactor has been discovered. Increasing the angle, the residence time decreases down to a minimum at 45° and then goes up for greater angles, as it can be seen in experiments for air/water.

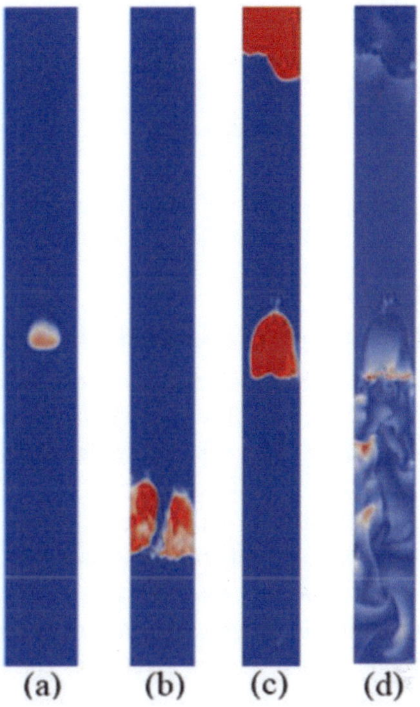

Fig. 5. Volume fraction 10mm (a), 30mm (b, c) and relative velocities (d).

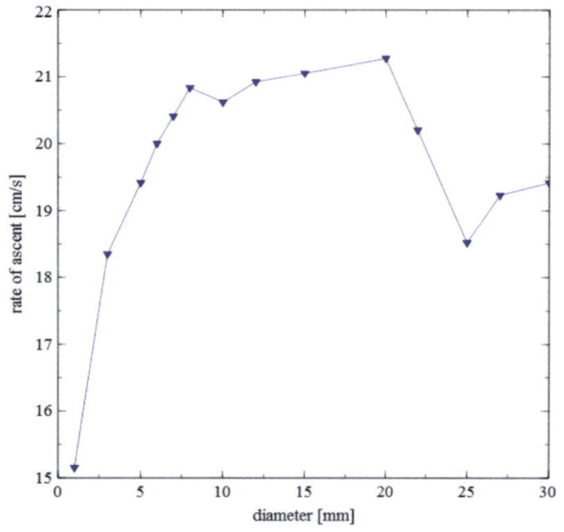

Fig. 6. Rates of ascent for different bubble sizes.

Magnetohydrodynamic flow control

To influence bubble residence times, one opportunity is given by the electromagnetic stirring and heating. This can be implemented by an inductor around the bubble column. To calculate the influence of an electromagnetic field on the bubble flow, additional relations, in detail described in [16], are necessary. The Lorentz force that is caused by the induced electromagnetic field of an inductor, has an influence on the velocity field of the melt. This influence and the coupling between hydrodynamic and electromagnetic effects can be described by the following equations:

$$\nabla^2 \vec{B} / (\mu_{mag} \sigma_{el}) + (\vec{B} \cdot \nabla)\vec{v} - (\vec{v} \cdot \nabla)\vec{B} = 0 \quad (25)$$

$$\vec{j} = \sigma_{el} \left(-\nabla \phi + [\vec{v} \times \vec{B}] \right) \quad (26)$$

$$\nabla \cdot \vec{v} = 0 \quad (27)$$

$$\partial v / \partial t + (\vec{v} \cdot \nabla)\vec{v} = -\nabla p / \rho_l + \nu_l \nabla^2 v + \vec{f} \quad (28)$$

Here $\vec{B}$ describes the magnetic field induction, $\vec{j}$ is the current density, μ_{mag} the magnetic constant, σ_{el} the electrical conductivity of the liquid, φ the electrical potential. The vector $\vec{f}$ describes the volume forces that include gravity g and Lorentz force $\vec{f_L} = \vec{j} \times \vec{B}$. In this case an additional force vector $\vec{F}$ appears in the momentum equation described in section 4.1.

The Lorentz forces generate whirl structures in the melt with velocity maxima near the reactor wall (see figure 4.4). As in many melting processes, electromagnetic stirring can be

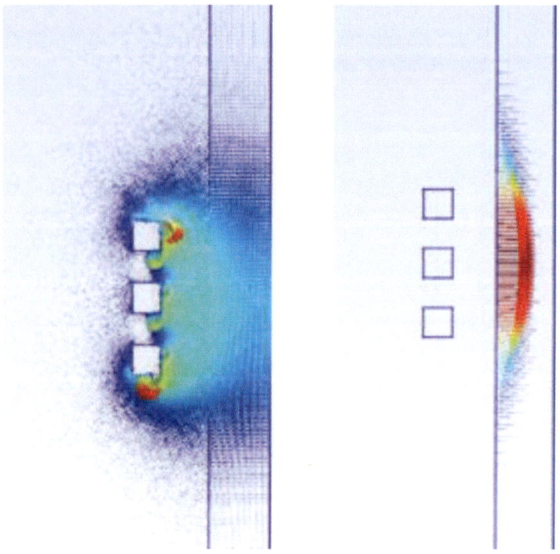

Fig. 7. Magnetic flux density (left) and resulting Lorentz-forces (right).

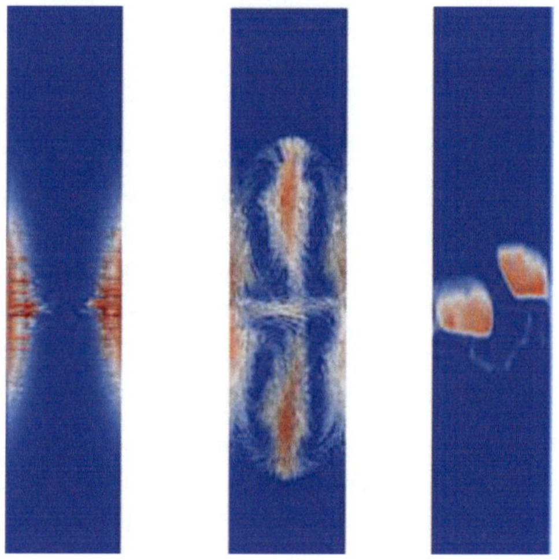

Fig. 8. Lorentz-force (left), melt velocity field after 5s (middle) and split bubble (right).

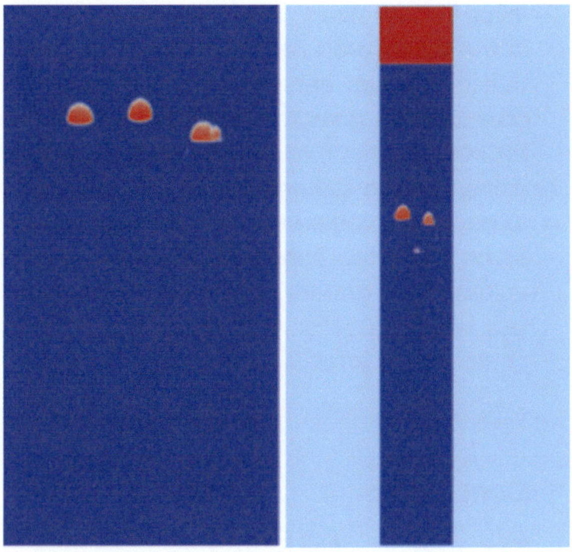

Fig. 9. Volume fractions for 3 inlets for higher (left) and dR/dB with resulting coalescence (right).

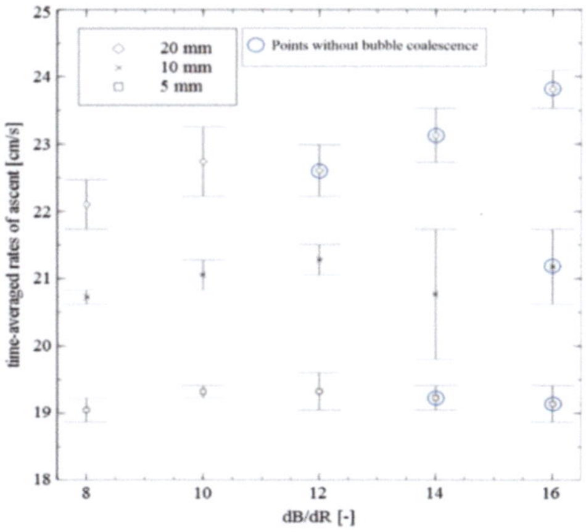

Fig. 10. Rates of ascents and coalescence smaller points for 3 inlets.

achieved, depending on the inductor frequency and the corresponding skin depth. To optimize the bubble dynamics and the resulting hydrogen yield, the idea is to use the turbulent shear force of the induced melt velocities to split bubbles into blebs. According to result in section 4.2., especially for bubble diameters around 7 mm, a diameter reduction results in higher residence times, besides the surface for heat transfer and reaction increases.

In former studies a volume of fluid method and a k-ω-SST-turbulence model has been implemented in the commercial tool *Ansys.* To discover the influence of an electromagnetic field on the bubbles, a coil with three windings and a width of 10 mm each is implemented in the lower part of the reactor using frequencies of 1 kHz and currents of 1 kA. The field has been switched on before the bubbles were inserted to get the necessary whirl structures. Former simulations in Ansys Fluent have shown that bubbles with diameters bigger than 6 mm have been split up. For greater bubble diameters around 9 mm an increase of residence time has been shown. In actual investigations a solver for dynamical coupling the Eulerian method with electromagnetics has been developed in OpenFOAM. The electromagnetic field and the resulting Lorentz forces are calculated via the open source tool *getDP* (figure 7) and-coupled with Eulerian method in *OpenFOAM* (figure 8). In actual studies inductor current and frequencies have been varied. A higher skin depth respectively a decreased the frequency, will increase the stirring effect due to modify distribution of the Lorentz forces. Bubble splitting can be better achieved when decreasing the skin depth by increasing the inductor frequency. But if the current is not changed, a higher power lever would also be needed. Depending on the relation of reactor diameter and skin depth, energy losses reduces with higher frequency which can be used for splitting the bubbles. Generally, it has to be noticed that the change of coupling between the Lorentz forces and the multiphase flow cannot be done with every time step. To identify such phenomena clearly also additional validations using neutron radiography are planned.

Multi-inlet coalescence studies

For a possible scaling up to industrial scale it is interesting to use more than one inlet in the reactor to increase the efficiency. In previous studies, a porous sparger has been used as an alternative inlet. The problem that is numerical described in [17], is a strong coalescence effects of the resulting bubbles above the porous media. To avoid this, the following passage will include different studies for the relation between the characteristic lengths of the flow problem. This has also been done with regards to a possible scaling up. In these studies, three inlets were discovered. For this case the initial distances between rising bubbles as well as reactor diameter were varied. Decreasing the reactor diameter and the initial distances of the 3 bubbles down to a critical point leads to coalescence (figure 9). Here the reactor volume (dark blue area in figure 9) is reduced. Detailed results are shown in figure 10. Here dR describes the characteristic length of the continuous phase (reactor diameter), while dB can be described as the characteristic length of the dispersed phase (initial bubble diameter). In the diagram the points without bubble coalescence are marked with blue rings. A range for the rates of ascents for the three bubbles is marked. It can be seen that especially for 10 mm bubbles a greater distance between the bubbles is necessary and for 20 mm also smaller distances avoid coalescence effects. Especially for 5 mm bubbles, an avoidance of coalescence leads to higher residence times. For 10 mm no strong influence can be seen while for 20 mm bubbles residence times even decrease without coalescence. Generally, especially the middle of the three bubbles is influenced by the other ones, and also wobbling effects appear. As seen in section 4.2, an increase of bubble size on smaller bubbles, e.g. by coalescence, has stronger influence on the rates of ascents.

In these studies, the distance between the bubbles are equally separated to the reactor, so the relations dR/dB can be recalculated to initial bubble distances to bubble diameter (aB/dR). So dR/dB = 12 equalsaB/dR = 2, dR/dB = 14 equals aB/dR = 2,5 and dR/dB=16 equals aB/dR = 3. For the case dR/dB = 16 (aB/dR = 3) in every discovered cases no coalescence appeared. In other words, the initial bubble distance has to be 3 times higher than initial bubble size to avoid bubble coalescence in the discovered range. It has to be noticed, that the zigzag behavior of the bubbles is increased with greater volume flows, which can be seen in simulations and also experiments using neutron radiography. If it is not possible to generate single bubbles with approximately same diameter, especially at higher constant volume flow, greater distances between the inlets would necessary to avoid the described coalescence effects.

Conclusion

The main objective of the presented work was to better understand the influence of some major influencing parameters on the direct thermal methane cracking within a liquid metal bubble column reactor. The feasibility of the method based on liquid metal technology was successfully demonstrated and presented in [2-4].

The results of a mechanistic thermo-chemical 1D-Model have confirmed the strong influence of liquid metal temperature on the methane cracking process. The model predictions are in good agreement with the experimental results. The developed model can be applied to investigate the influence of operating parameters and gas residence time in order to improve the process efficiency.

In the numerical investigation, dynamic studies for different bubble diameters were done. The simulations of volume flows as used in experiments show slug flow dynamics at highest volume flows. Generated bubble sizes vary with volume flow, but can be dispersed while dwelling in the reactor. Generally increasing the volume flow results in stronger bubble wobbling and appearance of strong swirls for higher volume flows. Another aspect, the splitting of methane bubbles into blebs via electromagnetic stirring was also explored. It has been shown that for bubbles with a sufficient high diameter bubbles can be split. Especially for greater bubbles around 9 mm, this split also result in greater residence times, which potentially lead to higher hydrogen yield. Instead of splitting the bubbles, it is generally interesting to avoid their coalescence. Several multi-inlets with ranges of bubble residence times as well as coalescence points were numerically discovered with regards to a possible scaling up of the system. The points of coalescence vary with bubble diameters and the relation between bubble diameter and reactor diameter.

The next steps in technological development will be scaling-up to prototype level and further investigations in view of increasing methane conversion efficiency. The possibility of efficient continuous carbon removal from the reactor may be the key factor for the practical implementation of the developed technology.

Acknowledgement

This study was financially supported by the Helmholtz Society within the Helmholtz Alliance for Liquid Metal Technologies (LIMTECH).

The authors would like to sincerely thank Prof. Carlo Rubbia, Prof. Alberto Abànades, Dr. Delia Salmieri and Dr. Stefan Stückrad of the Institute of Advanced Sustainability Studies, Potsdam, who piloted and enabled a previous complementary project on the topic and supported this project in multiple ways.

<u>Partner</u>

We warmly thank Tobias Geißler, Tristan Fehling and Egbert Baake from Leibniz University of Hannover, Institute of Electrotechnology for the greater encouragement and effective cooperation. We are grateful for your big contribution that you included to this project. Without your help and support, the project would not have been successful.

References

[1] Heinzel, A.; et al.; *Liquid Metals as Efficient High-Temperature Heat-Transport Fluids; Energy Technology.* Vol. 5, p1026 – 1036, 2017.

[2] Geißler, T.; *Methanpyrolyse in einem Flüssigmetall – Blasensäulenreaktor.* (Karlsruhe: Karlsruher Institute of Technology) 2017

[3] Geißler, T., et al.; *Experimental investigation and thermo-chemical modeling of methane pyrolysis in a liquid metal bubble column reactor with a packed bed; Hydrogen Energy.* Vol. 40, p 14134 – 14146, 2015.

[4] Abánades A.; et al.; *Development of methane decarbonisation based on liquid metal technology for CO2-free production of hydrogen. Hydrogen Energy; Hydrogen Energy*; Vol. 41, p 8159–8167, 2016.

[5] Geißler, T.; et al.; *Hydrogen production via methane pyrolysis in a liquid metal bubble column reactor with a packed bed; Chemical Engineering.* Vol. 299, p 192–200, 2016.

[6] Zhang, Y.; et al.; *Colloids and Surfaces A: Physicochemical and Engineering Aspects.* Vol. 223, p 45–54, 2003.

[7] Alchagirov, B. B.; et al.; *Surface tension of tin and its alloys with lead; Russian Journal of Physical Chemistry.* Vol. 81, p 1281-1284, 2007.

[8] Alchagirov, B. B. and Chochaeva, A M.; *Temperature dependence of the density of liquid tin;* 2000 *High Temperature.* Vol. 38, p 44-48, 2000.

[9] Melissari, B. and Argyropoulos, S. A.; *Development of a heat transfer dimensionless correlation for spheres immersed in a wide range of Prandtl number fluids; Heat and Mass Transfer.* Vol. 48, p 4333-4341, 2005.

[10] Hirt, C. W. and Nichols, B. D.; *Volume of fluid (VOF) method for the dynamics of free*

boundaries; Journal of Computational Physics. Vol. 39, p 201-225, 1981.

[11] Drew, D. A.; *Mathematical Modeling of Two-Phase Flow; Ann. Review Fluid Mech.* Vol. 15, p 261-291, 1983.

[12] Drew, D. A. and Lahey R. T.; Analytical modelnig of multi phase flow; *Particulate Two-Phase Flow* (Butterworth-Heinemann, Boston), p 509-566, 1993.

[13] Schiller, L. and Naumann, Z.; *A Drag Coefficient Correlation; VDI Zeitung.* Vol. 77, p. 318-320, 1935.

[14] Ranz, W. E. and Marshall, W. R.; *Evaporation von Drops.Part I; Chem. Eng. Progr.* (Madison), Vol. 48, p. 141-146, 1952.
Ranz, W. E. and Marshall, W. R.; *Evaporation von Drops.Part IIChem. Eng. Progr.* (Madison*),* Vol. 48, p. 173-180, 1952.

[15] Behzadi, A.; et al.; *Modelling of dispersed bubble and droplet flow at high phase fractions; Chemical Engineering Science.* Vol. 59, p. 759-770, 2004.

[16] Tucs, A.; et al.; *Implementation and verification of numerical model for gas bubble dynamics in electroconductive fluid; Proc. 11th Int. Conf. on Numerical Analysis and Applied Mathematics.*(Rhodes*)* p 21-27, 2013.

[17] Fehling, T.; et al.; *Numerical modeling of bubble dynamics in liquid metal exposed to electromagnetic fields for hydrogen production; International Journal of Applied Electromagnetics and Mechanics*. Vol. 53, p. 111-120, 2017.

Group: Accident Management Systems

Smart Grid Resilience: Concepts on dealing with Power Scarcity for decentralized Power Systems

Sadeeb Simon Ottenburger, Wolfgang Raskob

Introduction

A better understanding of upcoming technological transformations and their impact on critical infrastructure systems and security of supply is one of the main drivers of our research. The transformation of the classical power system into a smart decentralized power system is one of the most prominent and societally relevant examples of such transformations - the ongoing increase of automation and power consumption illustrates it's increasing importance accompanied by a simultaneously increase of vulnerabilities. Furthermore, a drastic change of power consumption, e.g. by an increased usage of electric vehicles, may result in unforeseen loads that cannot be managed by the utility provider e.g. in terms of demand side management. Therefore, our research deals with the following topics: Assessing the impact of different power and ICT- network structures on the resilience of urban systems. Development of new risk-based power distribution mechanisms dealing with power scarcity in order to avoid large-scale blackouts and improve security of supply (Ottenburger et al. 2018b; Ottenburger und Münzberg T. 2017). The methods that are used ground on modelling various critical infrastructures, power-, and ICT-infrastructures separately, but also new resilience measures. The key idea is to consider smart grid topology and power distribution mechanisms as model parameters. Thus, by varying these parameters for a specific region, e.g. an urban area, good structures and distribution mechanisms may be identified.

Urban Resilience and Power System

The concept of urban resilience encompasses various types of resilience dimensions such as the social, economic or physical infrastructure dimension (Cimellaro 2016). Critical infrastructure (CI) services such as the supply of electricity, drinking water, and health care provide vital services for the population, thus disruptions or failures of these services are hazardous and can lead to injuries or even losses of life, property damages, social and economic disruptions or environmental degradations. Therefore, CIs constitute a pivotal aspect in urban resilience considerations and establishing and implementing sophisticated continuity management concepts w.r.t. CIs can be regarded as one of the major building blocks for preserving or enhancing urban resilience. Most of the CIs like water supply, hospitals, pharmacies, and traffic- and transport systems rely on electricity - the circumstance of massive dependencies of other CIs to the electrical power entitles the electrical power grid to be considered as a *high ranked* CI. The generation and supply of electricity is currently about to undergo a fundamental transition (Farhangi 2010; Gungor et al. 2013). Due to the integration of smart meters, the consumers in the classical sense will have the eligibility to consume, produce and distribute electricity. The therefor necessary smart meters are electronic devices that monitor electricity consumptions and generations and allow two-way communications with other meters (Parhizi et al. 2015). However, to keep a stable electricity supply it is important that in-feed and consumption form an equilibrium. A smart grid construed as a complex and highly automated power distribution grid fundamentally relies on a rigorous multi-layered distribution management system in order to maintain grid performance and reliability. The architecture of a distribution management system allows a partitioning into several locally arranged and interconnected operation centers which themselves may be considered as local distribution management systems. A precise and secure operation of an energy management system of a smart grid, that operates - due to grid stability issues - automatically in real-time, heavily depends on the degree of accuracy of the transmitted quantities of interest.

For reasons of simplicity, in the remainder of this report infrastructures like companies, commercial buildings or households are included in the term CI.

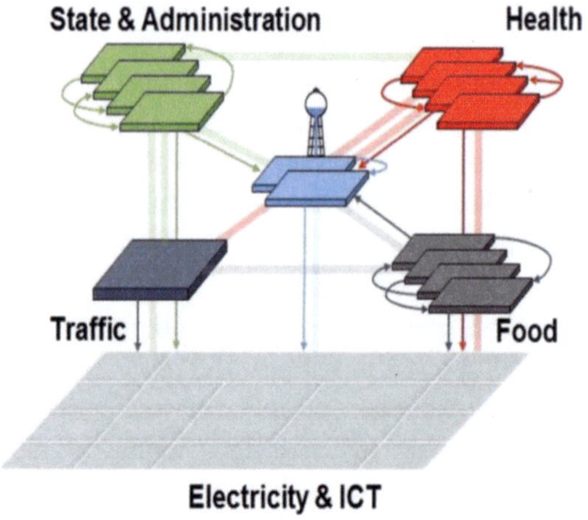

Fig. 1 Interdependent Critical Infrastructures

Vulnerabilities and Challenges of Future Power Systems

Many components of a smart grid are located not on the utility's premise and are therefore prone to physical damage. Since information technology systems are relatively short lived it is quite likely that outdated devices are still in service e.g. anti-virus software may be deprecated or hardware components may not comply with the latest requirements. Furthermore, the great number of devices are potential entry points for malicious cyber-attacks like malware spreading which may infect smart meters or other devices and can add or replace functions and disseminates, injecting false information by faking sensitive smart meter that can cause wrong decisions, and Denial-of-Service attacks by manipulating IP protocols that can delay, block or corrupt the transmission of information in order to make SG resources unavailable (Aloul et al. 2012).

Besides the increase of susceptibility on the cyber-physical component level, and despite the current and upcoming advancements and progresses in cyber security technologies, the vulnerability of the future power system itself and hence the probability of CI systems suffering from power outage grow, if there won't exist appropriate power continuity management strategies w.r.t. CIs. Another issue that should not be underestimated are drastic increases in power consumption that cannot be handled by demand side management techniques solely, e.g. e-vehicles and charging behavior.

Our research is pursuing the following objective: Developing new approaches to operationalize risk-based concepts to foster the developments of new strategies for CI protection by exploiting topological design options for smart grids and by proposing a risk-based smart grid control mechanism that is embedded in an energy management system.

Criticality and Degrees of Freedom

In view of net neutrality, this work perceives criticality as a risk-comparing framework, which we split up into different subcategories. The type of the power shortage scenario, the different relevancies of CIs and the timely varying demands for critical services determine the possibly timely varying so-called *global criticality* of CIs in an urban area. *Global criticality* of a CI can be considered as a function depending on relevancies of all other CIs, global system variables and the CI's *local criticality*, where *local criticality* is a function describing the critical state of a CI, reflecting its internal state, current and expected demand, current and expected fulfillment of demand etc. *Global criticality* and *local criticality* are of course power shortage scenario dependent. In the case of normal performance states of CIs and no occurring or expected power shortage, *global criticality* reflects the different relevancies of CIs compared to each other - these values are called *initial criticality* of CIs - for more details we refer to (Ottenburger et al. 2018b; Münzberg und Ottenburger 2018).

Operationalizing criticality should start in the phase of designing grid extensions or developments both from the ICT- but also from the physical power infrastructure perspective. This work especially focuses on two topological degrees of freedom in the design of smart grids. One topological degree of freedom refers to decomposition of a smart grid into so-called microgrids which may be disconnected from the overall smart grid and operate autonomously in island mode. A smart grid subdivided into microgrids has the potential to restrict cascading effects and hence to be less vulnerable against disruptions. Cascading effects due to dysfunctionalities of certain components or propagation of malware throughout a smart grid might be prevented by disconnecting the affected microgrids from the overall smart grid. Although having isolated disturbed or dysfunctional microgrids from the smart grid of a city, CI dependencies may cause issues in other parts and reduce the resilience of the city as a whole. Another topological degree of freedom refers to different configuration options w.r.t. smart grid components, e.g. overlaying network structures to provide redundancies within a microgrid.

Obviously, the network topology of a smart grid has significant effects on urban resilience particularly referring to the adequate provision of vital services of CIs. Taking *initial criticality* into account during the smart grid design phase can be regarded as a first proactive measure in the sense of preparedness. The rationale of applying *initial criticality* could be to distribute CIs with high *initial criticality* on different microgrids, avoiding a concentration of highly relevant CIs in one microgrid, or to build redundancies for CIs with high *initial criticality*. An elaborated topology of smart grids increases urban resilience.

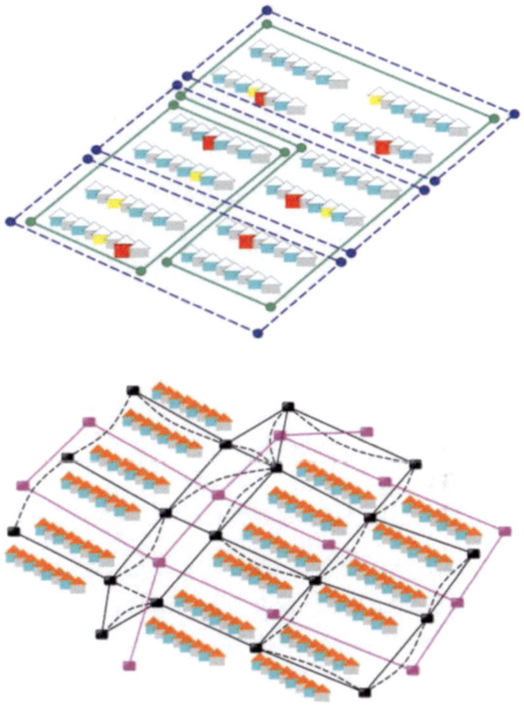

Fig. 2 Topological degrees of freedom: decomposition into microgrids and configuration of components

Resilience Measure: Supply Index

Advanced Metering Infrastructures (AMIs) including smart meters allow fine-grained power distribution management strategies that go beyond the classical strategies like rolling blackouts. A main task is to develop resilient and fair power distribution strategies in times of power shortage by exploiting the advantages of an AMI and smart meters. Therefore, new resilience metrics, measuring security of supply that complement known measures like SAIDI (System Average Interruption Duration Index) are developed.

Smart meter can be considered as an interface between a prosumer, e.g. a household, and the outer smart grid structure. An AMI that utilizes advanced smart meter technologies, allowing complex communication with distribution management system entities, enables a CI to transfer its current *local criticality* into the distribution management system. The distribution management system, being aware of all the *initial criticality* values, the relevancies of all other CIs, global system variables etc., is able to compute the current *global criticality* for each CI (Ottenburger et al. 2018b). In the case of a power shortage, caused by dark doldrums or cyber-attacks, a system knowledge about a *global criticality* distribution can be used for identifying optimal power distribution mechanisms. Therefore, a so-called Supply Index (SI) is applied: Let $i \in I$ be an infrastructure, and $c_i \in [0,1]$ the global criticality of i:

$\boldsymbol{SI} = \sum_{i \in I} \tilde{c_i}\, q_i(SP_i)$, where $\tilde{c_i} = \frac{c_i}{\sum_{j \in I} c_j}$, is the weighted global criticality and q_i a certain linear function measuring the quality of supply (Ottenburger et al. 2018a).

The spectrum of optimal power distribution mechanisms that are applicable strongly depends on the topology of smart grids or in other words smart grid topology determines the range of possible optimal power distribution techniques and massively influences urban resilience. During a power shortage, optimal power flows or power distribution policies should target at distributing electricity in such a way that the severity of the impact of a possible decrease of the overall performance of all CIs in an urban area is *minimum*. In enhanced power distribution policies, where *global criticality* is applied, rolling black outs in terms of dynamically connecting different microgrids with each other, might also be an option.

Results

The Smart Grid Resilience Framework

In an ongoing interdisciplinary research project (Raskob et al. 2015) UNF is collaborating with the Institute for Program Structures and Data Organization (IPD), the Institute for Industrial Production (IIP) and the Institute for Automation and Applied Informatics (IAI). The aim of this project is to develop a framework for simulating CIs against certain disruption scenarios and to perform systemic resilience assessments of regions of urban scale by mainly focusing on power shortage scenarios. Power scarcity may result from severe weather events, cyber-attacks, exceptional dark doldrums, or extreme loads.

A main feature of this framework is to use network topologies and power flow algorithms as model parameters that can be varied. The framework has a modularized structure: There is a power module, an ICT-module, and a CI-module, that simulates various CIs from the health sector and the water supply (Raskob et al. 2015; Münzberg T. et al. 2018).

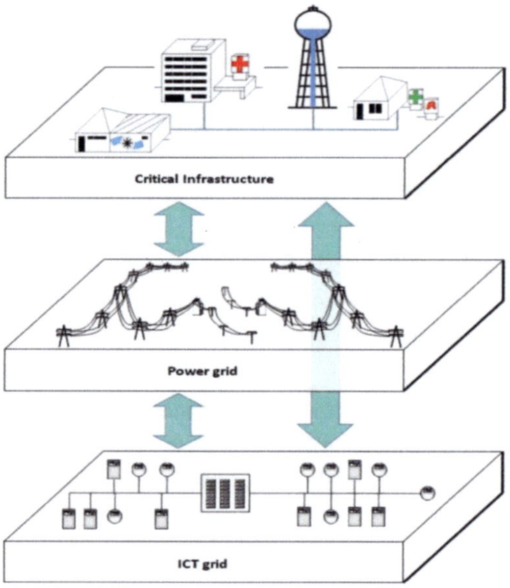

Fig. 3 Modular Structure of the Smart Grid Resilience Framework

In each of these modules, different disruptions/damages of infrastructures can be modelled separately and adjusted within a generic disruption framework. By applying the SI, the resilience of physical and ICT-network structures and power distribution strategies can be assessed. The nature of this framework is quite generic and could principally be applied to any urban area. Currently, the medium voltage distribution grid of Karlsruhe is modelled as the base physical power infrastructures upon which grid extensions models, as possible future scenarios, are further developed.

Optimal Power Flow applying *SI* and evolutionary Algorithms

As mentioned in the previous section, the Smart Grid Resilience Framework uses network topologies and power flow algorithms as model parameters that can be varied. First concepts on resilient power flow mechanisms were developed:

An infrastructure i may possess process flexibility or coping capacities that allow to specify a *power demand flexibility interval* $[P^i_{D,min}, P^i_{D,max}]$, where $P^i_{D,max}$ is the power demand for *normal process* mode and $P^i_{D,min}$ the power demand for at least running *some essential sub-processes.* There might be an infrastructure i that has criticality equal to 1 - in this case i has no power flexibility and thus demands a certain power value P^i_D.

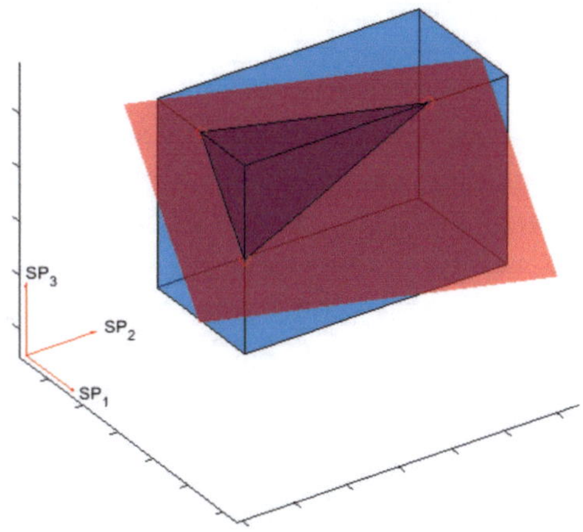

Fig. 5 75 % - scenario-hyperplane: power value vectors within the truncated power demand cube are potential solutions of the resilient optimal power flow

In the case of power shortage, e.g. only 75% of the normal power demand of all infrastructures is available, new resilient power distribution strategies in the energy management system of a smart grid are thinkable in order to avoid total blackouts - nevertheless, new distribution algorithms should not violate non-discrimination.

Before applying these new concepts on our distribution grid model of Karlsruhe, they are

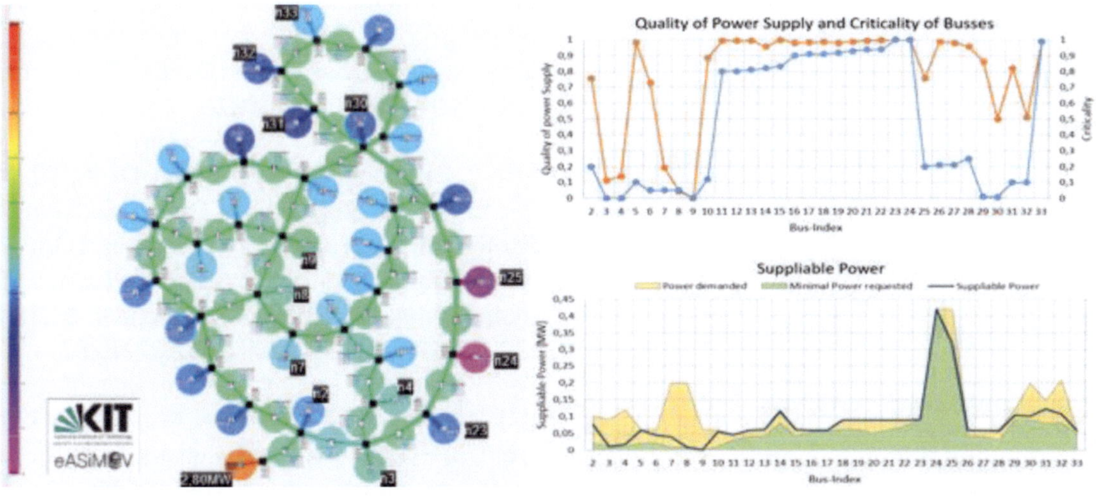

Fig. 4 First results of criticality based optimal power flows on the IEEE33 bus system for stationary load states

validated against simpler grid models e.g. standard IEEE-models:

For stationary load states optimal power flows on the IEEE33 bus system were calculated in a joint project with IAI - genetic algorithms were applied, in order to maximize SI while respecting fair distribution (Ottenburger et al. 2018a). Conclusively, the way we applied genetic algorithms can be considered as a possibility to specify the corresponding parameter in the Smart Grid Resilience Framework.

Summary

The ongoing transition of the power distribution system towards Smart Grids bears the chance to conceptually integrate principles of crisis prevention and management into power control mechanisms. Following this aspiration global criticality considered as a dynamic feature of CIs is a promising attribute that helps to bridge Smart Grids resilience and urban resilience in a sensible way. In cases of power shortages, where first, secondary, and tertiary controls are not able to stabilize the whole grid, global criticality as a further criterion can be applied for controlling the power flow in a Smart Grid in an urban resilient way and thus contribute to enhancing urban Continuity Management (CM).

The proposed simulation framework, allowing the variation of parameters like Smart Grid topology and power distribution mechanisms, can be applied to specific urban systems. Simulation studies against power disruption scenarios that are considered to be plausible for that particular urban system can be conducted. The outcomes of these studies would be

- an analytical view on Smart Grid infrastructure planning in terms of a selection of urban resilient Smart Grid design options and
- appropriate power distribution strategies or algorithms to deal with power shortage scenarios; these strategies or algorithms could be implemented in the Energy Management System (EMS) of a Smart Grid.

References

[1] Adrot, A.; Fiedrich, F.; Lotter, A.; Münzberg, T.; Rigaud, E.; Wiens, M.; Raskob, W. and Schultmann, F.; (2018): Challenges in Establishing Cross-Border Resilience, In: *Fekete A., Fiedrich F. (eds) Urban Disaster Resilience and Security. The Urban Book Series. Springer, Cham.*

[2] Aloul, Fadi; Al-Ali, A. R.; Al-Dalky, Rami; Al-Mardini, Mamoun; El-Hajj, Wassim (2012): Smart Grid Security. Threats, Vulnerabilities and Solutions. In: *International Journal of Smart Grid and Clean Energy*, S. 1–6. DOI: 10.12720/sgce.1.1.1-6.

[3] Bai, S.; Staudt, C.; Kaiser J.C.; Raskob, W. (2018): Knowledge Database and Regionalization of JRODOS for the HARMONE Project. In: *4th NERIS Workshop proceedings.*

[4] Cimellaro, G. P. (Hg.) (2016): Urban Resilience for Emergency Response and Recovery (Vol. 41): Cham: Springer International Publishing.

[5] Farhangi, H. (2010): The path of the smart grid. In: *IEEE Power and Energy Mag.* 8 (1), S. 18–28. DOI: 10.1109/MPE.2009.934876.

[6] Gungor Cagri, V.; Sihan Dilan; Kocak, T.; Ergut, S.; Buccella, C.; Cecati, C.; Hancke, Gerhard P. (2013): A Survey on Smart Grid Potential Applications and Communication Requirements. In: *IEEE Trans. Ind. Inf.* 9 (1), S. 28–42. DOI: 10.1109/TII.2012.2218253.

[7] Kiwa, S.; Zheleznyak, M.; Boyko O.; Ievdin, I.; Pylypenko, O.; Mikhalsky, O. et al.: Updated module of radionuclide hydrological dispersion of the Decision Support System RODOS. In: *Geophysical Research Abstracts 20 (2018)* 2018.

[8] Möhrle, S.; Bai S.; Müller, T.; Munz, E.; Trybushnyi, D.; Raskob, W.: Web-based decision support system for emergency management - System architecture and enhancement possibilities. In: *4th NERIS Workshop "Adapting nuclear and radiological emergency preparedness, response and recovery to a changing world", Dublin, Ireland,* 2018.

[9] Möhrle, S.; Bai S.; Müller, T.; Munz, E.; Trybushnyi, D.; Raskob, W:Triggering Events and Distributed Responsibilities, Capabilities of Web-based Decision Support in Nuclear Emergency Management, 4th NERIS Workshop proceedings, to be published.

[10] Münzberg, T.; Ottenburger, S. S.: Schutz Kritischer Infrastrukturen: Kritikalität als Entscheidungsmaß zur Abwehr von Gefahr am Beispiel Stromausfall. In: *Was heißt Kritikalität? Zu einem Schlüsselbegriff der Debatte um Kritische Infrastrukturen. transcript Verlag* 2018.

[11] Münzberg, T.; Ottenburger, S. S. (Hg.) (2018): Schutz Kritischer Infrastrukturen: Kritikalität als Entscheidungsmaß zur Abwehr von Gefahr am Beispiel Stromausfall. Was heißt Kritikalität? Zu einem Schlüsselbegriff der Debatte um kritische Infrastrukturen. Bielefeld: transcript (Science Studies). Online verfügbar unter https://doi.org/10.14361/9783839442074.

[12] Münzberg, T.; Wiens, M.; Schultmann, F. (2017): A spatial-temporal vulnerability assessment to support the building of community resilience against power outage impacts. In: *Technological Forecasting and Social Change* 121, S. 99–118. DOI: 10.1016/j.techfore.2016.11.027.

[13] Münzberg T.; Müller T.; Raskob W. (2018): A Future-Oriented Agent-Based Simulation to Improve Urban Critical Infrastructure Resilience. In: *Fekete A., Fiedrich F. (eds) Urban Disaster Resilience and Security. The Urban Book Series. Springer, Cham.*

[14] Ottenburger, S.; Cakmak, H. K.; Jakob, W.; Blattmann A.; Deines E.; Trybushnyi, D. et al. (2018a): Smart Grid Resilience - Security of Supply 2.0. Poster - Jahrestagung Zentrum Energie.

[15] Ottenburger, S.; Münzberg T. (2017): An Approach for Analyzing the Impacts of Smart Grid Topologies on Critical Infrastructure Resilience. In: *ISCRAM 2017 Conference Proceedings - 14th International Conference on Information Systems for Crisis Response and Management.*

[16] Ottenburger, S. S.; Airaksinen, M.; Pinto-Seppa, I.; Raskob, W. (2017): Enhancing urban resilience via a real-time decision support system for smart cities. In: 2017 International Conference on Engineering, Technology and Innovation (ICE/ITMC). 2017 International Conference on Engineering, Technology and Innovation (ICE/ITMC). Funchal, 27.06.2017 - 29.06.2017: IEEE, S. 836–844.

[17] Ottenburger, S. S.; Bai S. (2017): Simulation Based Strategic Decision Making in Humanitarian Supply Chain Management. In: *4th International Conference on Information and Communication Technologies for Disaster Management (ICT-DM' 2017), Münster, Germany,*

[18] Ottenburger, S. S.; Daniell, J. (2018): Direct and Indirect Vulnerability Analysis Framework for Decentralized Power Systems. In: *International Conference on "Natural Hazards and Risks in a Changing World", Potsdam.*

[19] Ottenburger, S. S.; Münzberg, T.; Strittmatter, M. (2018b): Smart Grid Topologies Paving the Way for an Urban Resilient Continuity Management. In: *International Journal of Information Systems for Crisis Response and Management* 9 (4), S. 1–22. DOI: 10.4018/IJISCRAM.2017100101.

[20] Parhizi, Sina; Lotfi, Hossein; Khodaei, Amin; Bahramirad, Shay (2015): State of the Art in Research on Microgrids. A Review. In:

IEEE Access 3, S. 890–925. DOI: 10.1109/ACCESS.2015.2443119.

[21] Raskob, W. (2018): Emergency preparedness: advancement and still open gaps in Europe. In: *5th European IRPA Congress "Encouraging Sustainability in Radiation Protection", The Hague, NL,*

[22] Raskob, W.; Almahayni, T.; Beresford, N. (2018a): Radioecology in CONFIDENCE. Dealing with uncertainties relevant for decision making. In: *Journal of environmental radioactivity* 192, S. 399–404. DOI: 10.1016/j.jenvrad.2018.07.017.

[23] Raskob, W.; Bai S.; Staudt, C. (2018b): OPERRA-HARMONE, Knowledge Database and FDMT regionalization. In: *4th NERIS Workshop "Adapting nuclear and radiological emergency preparedness, response and recovery to a changing world", Dublin, Ireland,*

[24] Raskob, W.; Beresford, N. (2017): CONFIDENCE: Dealing with Uncertainties Relevant for Decision Making. In: *4th International Conference on Information Systems for Crisis Response and Management (ISCRAM 2017), Albi, F, May 21 - 24, 2017.*

[25] Raskob, W.; Bertsch, V.; Ruppert, M.; Strittmatter, M.; Happe, L.; Broadnax et al. (2015): Security of electricity supply in 2030. In: *Critical Infrastructure Protection and Resilience Europe (CIPRE), Conference & Expo, The Hague, the Netherlands.*

[26] Raskob, W.; Möhrle, S. (2017): Nuclear Emergency Response and Big Data Technologies. In: *Workshop on Big Data in SC5 Climate Action & Environment, Brussels, B, November 6, 2017.*

[27] Raskob, W.; Münzberg, T. (2017): Schmutzige Bomben und ihre möglichen Folgen für die Bevölkerung. In: *Crisis Prevention - Das Fachmagazin für Innere Sicherheit, Bevölkerungsschutz und Katastrophenhilfe 1*, S. 41–43.

[28] Raskob, W.; Schichtel, T.; Landman, C. (2017): Das Entscheidungshilfesystem RODOS zur Online- und Realtime-Unterstützung. In: *Behördenseminar "Notfallschutz außerhalb der Anlage", GRS Köln, 15.-16. März 2017.*

[29] Staudt, C.; Raskob, W. (2017): Modellierung des Radionuklidtransfers in einem europäischen Entscheidungshilfeprogramm. In: *2. Workshop - Helmholtz Cross Program Activity Querschnittsthema Strahlenforschung "Transportprozesse in Mensch und Umwelt", GSI Helmholtzzentrum für Schwerionenforschung, Darmstadt, 24.-25.10.2017.*

[30] Wiens, M.; Raskob, W.; Diehlmann, F.; Wandler, S.; Schultmann, F. (2018): Exposure and vulnerability of the energy system to internal and external effects. In: *Energy as a Sociotechnical Problem. Ed. by Christian Büscher, Jens Schippl, Patrick Sumpf. Routledge, London, pp. 96-122.*

[31] Zhang, X.; Raskob, W.; Landman, C.; Trybushnyi, D.; Haller, C.; Yuan, H. (2017): Automatic plume episode identification and cloud shine reconstruction method for ambient gamma dose rates during nuclear accidents. In: *Journal of environmental radioactivity* 178-179, S. 36–47. DOI: 10.1016/j.jenvrad.2017.07.014

Group: Hydrogen

Hydrogen Risk Assessment for Nuclear Applications and Safety of Hydrogen as an Energy Carrier

Paul Coupe, Andrei Denkevits, Nicolas van Dijk, Olaf Jedicke, Thomas Jordan, Birgit Kaup, Alexei Kotchourko, Mike Kuznetsov, Alexander Lelyakin, Frank Prestel, Maria Stassen, Jianjun Xiao, Zhanjie Xu, Han Zhang

Introduction

In the years 2017 and 2018 the Hydrogen Group continued to develop the in-house CFD HyCodes, GASFLOW and COM3D, and conducted several experimental programs to deepen the understanding of the transient flame acceleration and deflagration-detonation-transition or to provide model validations data. As cryogenic liquid hydrogen has a high potential for supporting the scaling up of applications of hydrogen as an energy carrier, these fundamental phenomena have to be studied also in the low temperature domain. The Hydrogen Group is coordinating the European pre-normative research project PRESLHY to this end. Besides the group has been active in the educational domain by teaching Hydrogen Technologies and managing a related European project NET-TOOLS.

In the application domain a safe catalytic cleaning device for high pressure hydrogen has been developed in the frame of the cooperation with the Instituto Technico Buenos Aires ITBA.

Pre-normative Research for the safe use of liquid hydrogen

For scaling up the hydrogen supply infrastructure the transport of liquefied hydrogen is the most effective option due to the energy density. Especially for the transport sector with the planned large bus fleets, the emerging hydrogen fueled train, boat and truck projects and even for the pre-cooled 70 MPa car fueling, liquid hydrogen (LH2) offers sufficient densities and gains in efficiency over gaseous transport, storage and supply. However, LH2 implies specific hazards and risks, which are very different from those associated with the relatively well-known compressed gaseous hydrogen. Although these specific issues are usually well reflected and managed in large-scale industry and aerospace applications of LH2, experience with LH2 in a distributed energy system is lacking. Transport and storage of LH2 in urban areas and the daily use by the untrained general public will require higher levels of safety provisions accounting for the very special properties. The quite different operational conditions compared with the industrial environment and therefore also different potential accident scenarios will put an emphasis on specific related phenomena which are still not well understood. Specific recommendations and harmonized performance based international standards are lacking for similar reasons.

The Hydrogen Group coordinates the project PRESLHY, which is supported by the European Fuel Cell and Hydrogen Joint Undertaking (FCH 2 JU). The consortium consisting of 10 European research institutions and industry representatives will do research for the most relevant and poorly understood phenomena related to high risk scenarios from 2018 until end of 2010. With the new knowledge generated by this research work science based and validated tools, which are required for hydrogen safety engineering and risk informed, performance based, LH2 specific, international standards will be developed.

Besides the actual coordination the Hydrogen Group of IKET is leading the combustion phenomena work package and conducts an essential part of the experimental program ranging from cryogenic gas dispersion, ignition phenomena to transient combustions phenomena.

So far a review of the state-of-the-art and existing regulations, codes and standards have been conducted. First screening experiments related to ignition by electro-static charge build-up in the accidental cold gas release have been conducted in the test cell Q160.

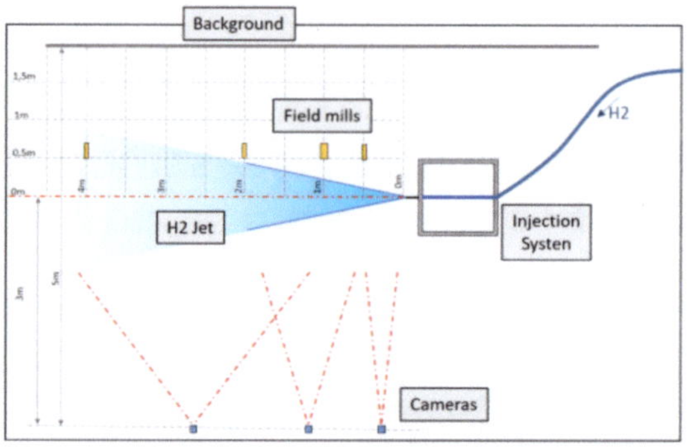

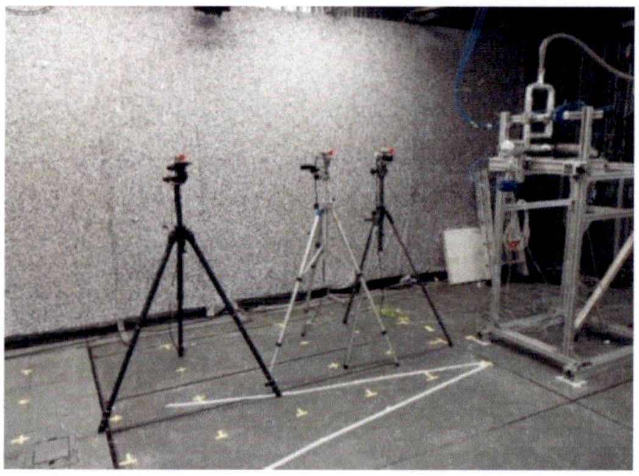

Fig. 19: Experimental set-up of for determination of electrostatic field build-up of cryogenic hydrogen jet releases in test cell Q160 (top: principle view from top experiments; bottom: photography with indicating positions of field mills by red dots)

Cryo-jet ignition by self generated electrostatic charges

The electrostatic field generated by the cryo-jet was measured with a set-up shown in Fig. 19. With the field mills an electrostatic field was identified, which seem to be induced by a cloud of charged particulates carried by the jet. It typically reaches values in the order of 1000 V/m. The particulates must be too small to be seen on the cameras.

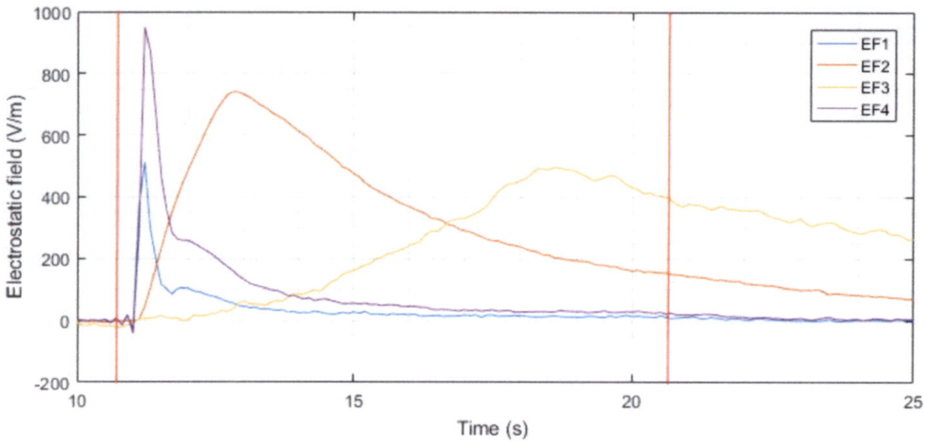

Fig. 18: Recordings of the different Electric Fieldmills (EF) with EF1 close to release point and EF3 at longest distance from release point; EF4 is special field mill with highest sensitivity close to EF1.

As we move further away from the nozzle, the peeks of electric field are delayed in time, and last longer, see Fig. 18. This can be explained by the decreasing velocity of the jet along the central axis. The further away from the nozzle, the longer it takes for the moving cloud to reach, and pass in front of the sensor. The delay and the duration of the peek reduces when the mass flux is increased.

After replacing hydrogen with helium, to exclude material specific issues, similar effects have been observed. Irrespective of the released gas, with sufficiently high ambient humidity and low temperature of the pre-cooled nozzle, ice crystals will form on the release nozzle, see Fig. 21. When the release is initiated the gas will tear of and entrain some small ice crystals. This process involving solid particles and strong shear flow forces is very likely the origin of the electrical charges observed.

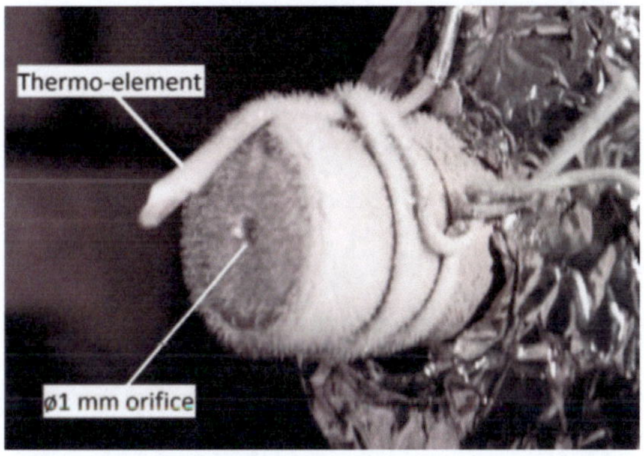

Fig. 21: Pre-cooled release nozzle with ice particles formed on

So, the results of these preliminary test might be summarized as:

- This phenomenon occurs only at cold temperatures of the released gases and with some minimum ambient humidity.
- Then jet carries electrical charges, which might be induced by ice particles formed at the cold nozzle before the actual release (possibly also during the release)

Whether the measured electrical fields are sufficiently strong and effecting the relevant premixed zones of the jet has to be clarified in the sub-sequent test program.

Characterisation of cryogenic hydrogen discharges

Discharge coefficients of cryogenic (T < 120K), high pressure (p < 20 MPa) hydrogen reservoirs will be determined in the DISCHA experiment series (see Fig. 22).

The experiments have been extended to simultaneously investigate the dispersion, ignition and combustion of the released cold hydrogen. This required the installation of an appropriate optical measurement system, consisting of several cameras. The main technique applied for the near field as well as for the far field observations of the cold gas jet behavior will be

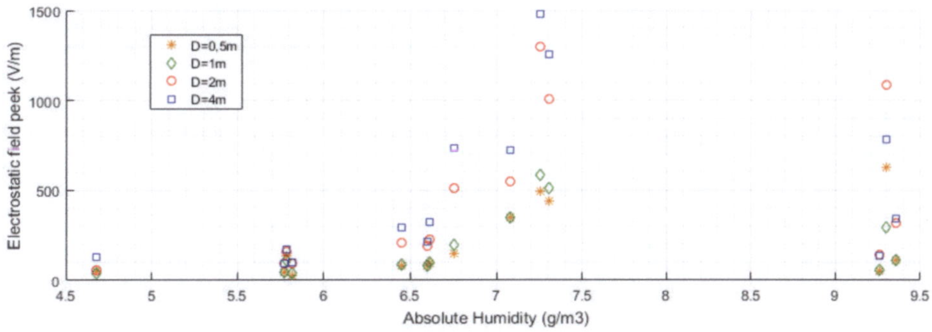

Fig. 20: Electric field strength dependent on ambient humidity and mass flux (release nozzle diameter, respectively)

Background Oriented Schlieren (BOS). So besides the actual objective to determine discharge coefficients for the cryogenic hydrogen, the experiments will provide indications on whether the quite simple BOS technology might be used for far field observation with natural backgrounds, like those encountered at hydrogen fueling stations, for instance.

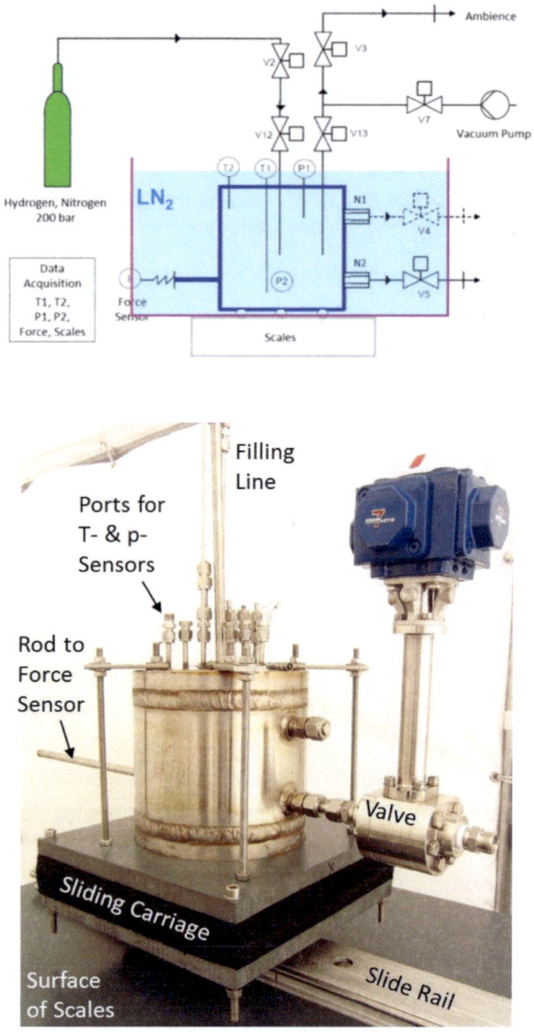

Fig. 22: DISCHA facility and experimental set-up for 80K hydrogen discharge releases

The minimum temperature which might be achieved by liquid nitrogen cooling of the DISCHA facility is 80 K. Therefore cold, but only single phase gaseous hydrogen releases might be studied with DISCHA. Assessing the effects of the liquid-gas interface during a blow-down and the determination of corresponding discharge coefficients requires a specially prepared cryo-vessel (see Fig. 23). The design of this experimental set-up has been finished, the construction of the special flange is on the way.

Fig. 23: Experimental set-up CRYO-VESSEL for two phase cryogenic hydrogen blow-downs

For the investigation of flame acceleration (FA) and deflagration-to-detonation (DDT) phenomena at cryogenic temperatures the special combustion tube CRYOTUBE with internal LN2 cooling has been designed and built (see Fig. 24).

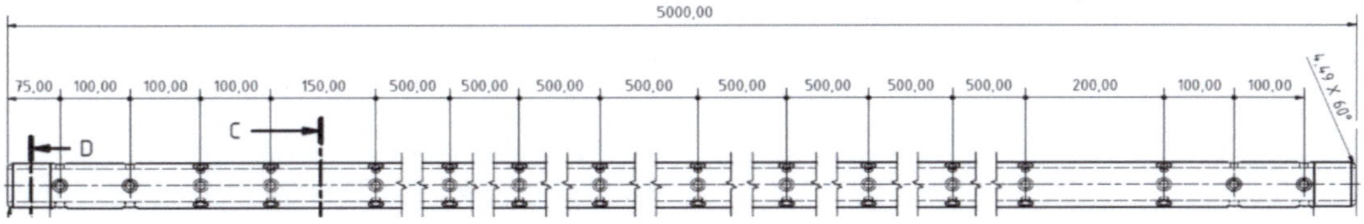

Fig. 24: LN2 cooled combustion tube CRYOTUBE with measurement ports

Results for the discharge, two phase blowdown and the cryogenic combustion characteristics are expected within the year 2019.

Catalytic Cleaning of High Pressure Hydrogen

As part of a cooperation with the Instituto Technico Buenos Aires, Argentina (ITBA) a catalytic cleaning process for high pressure electrolyzers has been investigated with a special experimental set up (see Fig. 25).

Several modifications for the feed gas arrangement have been used to achieve highest pressure in the experiments. Design criteria were derived by testing a few variants of the catalytic tube reactors (see Fig. 26), which were all equipped with the same small (about 4 mm in diameter) Pd-coated ceramic spheres, used as catalytic elements (see Fig. 27).

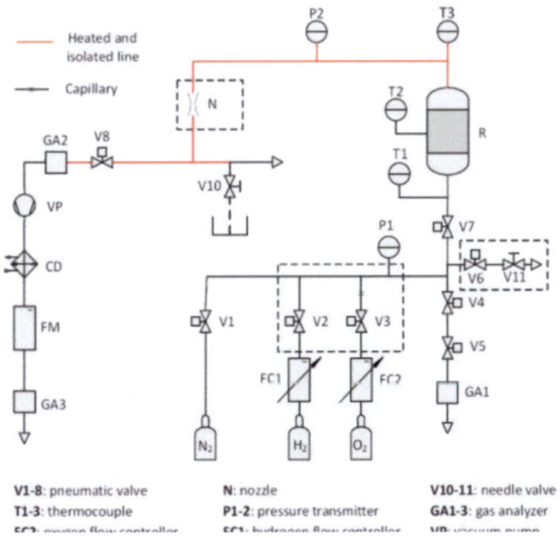

Fig. 25: PID of the experimental set-up for the catalytic cleaning of high pressure hydrogen

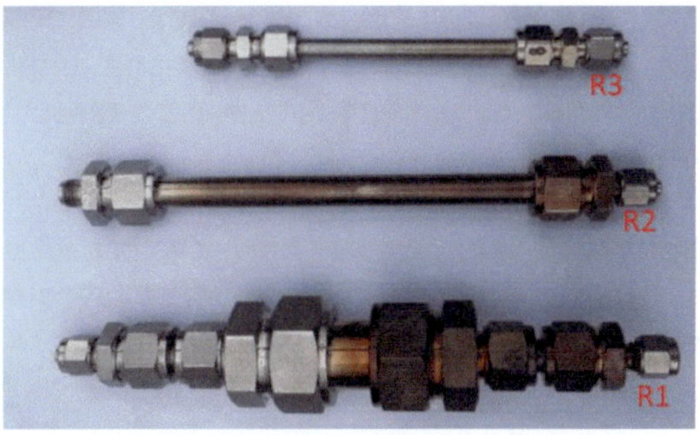

Fig. 26: Tested catalytic reactor variants

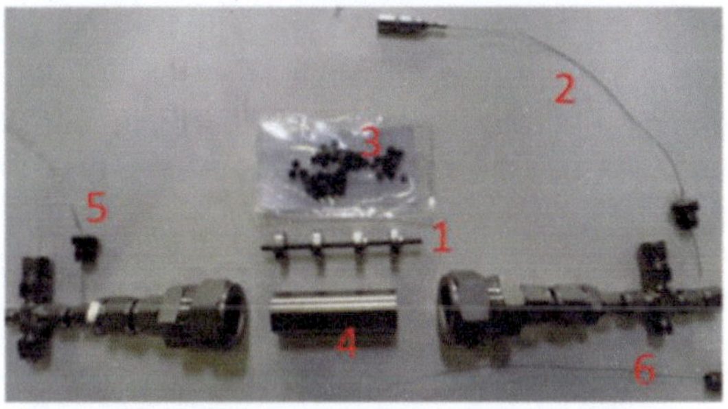

Fig. 27: Disassembled reactor variant R1 (with 1= spacer grid system, 2=exhaust thermocouple, 3=catalytic pellets,4=reactor shell, 5 = feed thermocouple, 6= reactor thermocouple)

Several test series were carried out with different set up configurations, most of them of them controlled with a stepwise increase of the oxygen contaminant concentration. The main results are the dry hydrogen or oxygen concentrations at the reactor outlet and the efficiency defined by comparing these values to the corresponding inlet concentrations.

The efficiencies of the catalytic reactors show inversely proportional correlation to the residence time of the gas mixture in the reactors. Feed concentration and reactor temperature show linear correlations. This outcome may be used to estimate the feed concentration by measuring the reactor temperature, being this a simple alternative instead of using gas analyzers.

The test performed with oxygen rich concentration shows a similar behavior when compared to hydrogen rich mixtures. However a feed concentration threshold in order the catalytic process to start is observed and further tests should be performed to better assess its behavior.

Pressure is found to have a positive impact on the efficiency of the process for the pressure range of 10 to 70 bar with a 10% efficiency increase. Tests performed at higher pressures (~ 90 bar) show a strange behavior leading to the assumption that steam adsorption is being developed by the catalytic material. Taking into account the aforementioned plus the fact that test pressure was limited due to needle valve choking events, modifications to the experimental set up are proposed in order to conduct tests at higher pressures.

The reactor variant R2 (tests 36, 37 and 57 in Fig. 28) turned out to be one with best performance under the conditions which are relevant for the prototypical alkaline high pressure electrolyser developed at ITBA. The reactor will be integrated there in 2019.

Overall, the catalytic cleaning process tested showed good performance and behavior understanding for the tests conducted at pressures below 70 bar. On the other hand, the working pressure limitation of the experimental set up prevented to fully understand the behavior of the process at pressures above 90 bar and further modifications to the facility should be conducted.

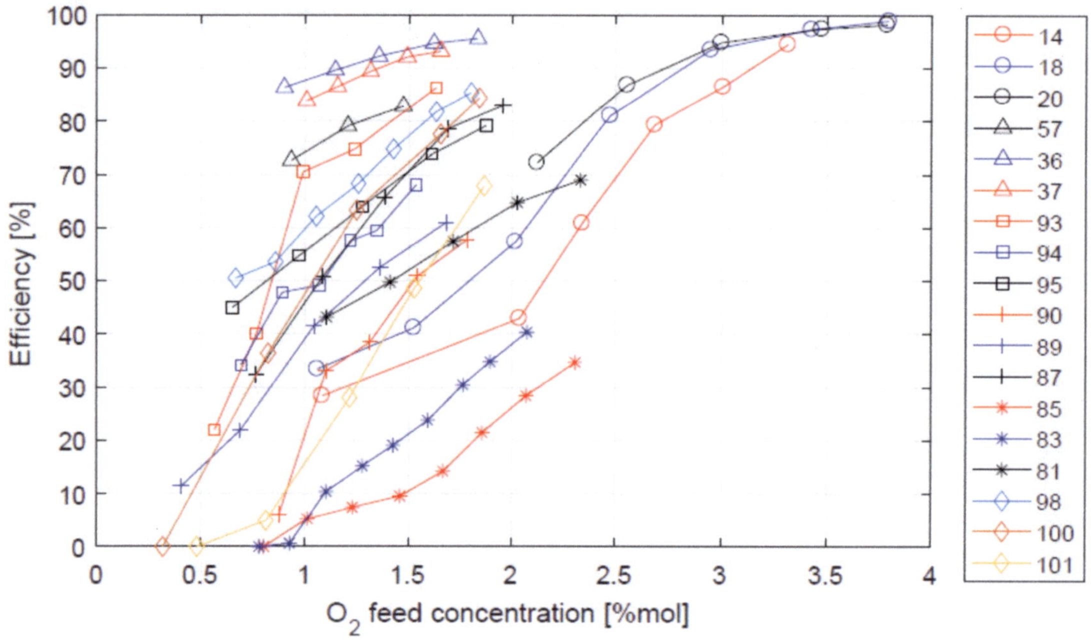

Fig. 28: Efficiency as a function of feed oxygen concentration

Gruppe: Energie- und Verfahrenstechnik

Messung der thermodynamischen Thermalwassereigenschaften an Geothermiestandorten

Elisabeth Schröder, Klaus Thomauske, Dietmar Kuhn

Einleitung

Der Energieinhalt und die Menge und des geförderten Thermalwassers sind entscheidende Parameter für die Auslegung und Dimensionierung geothermaler Anlagen. Neben der Fördertemperatur spielen die thermodynamischen Eigenschaften des Thermalwassers eine große Rolle. Diese weichen von den Daten reinen Wassers teilweise sehr stark ab, was durch hohe Salz- und Gasgehalte bedingt ist.

Zuverlässige Angaben zu den Stoffwerten sind in der Literatur kaum zu finden, da die Komplexität der Thermalwasserzusammensetzung bisher nur unzureichend abgebildet ist. Umfangreichere Datensätze existieren für Natriumchloridlösungen, allerdings werden Salinitäts-, Druck und Temperaturbereiche, wie sie in der Geothermie vorherrschen, nicht abgedeckt. Einige Untersuchungen zu Natriumchloridlösungen und Meerwasser sind in [1-9] zu finden. Des Weiteren gibt es keine Zuordnung der Stoffwerte zu den jeweiligen geologischen Horizonten, was im Hinblick auf die Vorhersagbarkeit der Thermalwassereigenschaften bei Planungsbeginn von Geothermieanlagen wünschenswert wäre.

Um die thermodynamischen Eigenschaften unter Anlagenbedingungen zu bestimmen, wurde am KIT ein Prüfstand entwickelt mit Hilfe dessen die spezifische, isobare Wärmekapazität, die Fluiddichte und die kinematische Viskosität im Druck- und Temperaturbereich bis 30 bar und 170 °C ermittelt werden können. Die ausführliche Beschreibung der Messtechnik ist in [10] und [11] zu finden.

Im Rahmen des vom BMWI geförderten Verbundprojektes „PETher" wurden zehn Geothermiestandorte beprobt. Dabei konnten die drei geothermisch relevanten Regionen (Molassebecken, Oberrheingraben, Norddeutsches Becken) abgedeckt werden.

Diese in-situ durchgeführten Messungen wurden durch systematische Laboruntersuchungen ergänzt. Nach umfangreicher Datenanalyse durch die Projektpartner GeoT und GTN konnten für die geothermisch relevanten Horizonte typische Thermalwasserzusammensetzungen identifiziert werden. Diese wurden im Labor als Modellwässer nachgebildet und deren Stoffwerte im Temperaturbereich von 20°C bis 150 °C ermittelt.

Basierend auf den in-situ Messungen und den Modellwasseranalysen stehen nun Kennkurven zur Verfügung, die, abhängig von der Bohrtiefe und des Horizontes, eine deutlich verbesserte Vorhersage der thermodynamischen Thermalwassereigenschaften erlauben.

Ergebnisse

Molassebecken

Nach Analyse der Tiefenwasserzusammensetzung des Oberjura Aquifers, wurden für das Molassebecken die Wässer in fünf Gruppen gefasst, [12]. In Gruppe I, dem nördlichen Beckenrand, ist der Wassertyp Ca-(Mg)-HCO_3 mit einem sehr geringen Mineralgehalt von 497 mg/l zu finden. Im zentralen Becken überwiegt mit Gruppe II der Typ Na-Ca-Mg-HCO_3 bei mittleren Salzgehalten von 588-689 mg/l. Tiefenwässer vom Typ Na-HCO_3-Cl (Gruppen III und IV) sind im Niederbayerischen und im

Nordöstlichen Becken mit Salinitäten von 1000 – 1500 mg/l zu finden, wobei bei letzterem der Chloridanteil dominierend ist. Das Südwestliche Becken, Gruppe V, hat den höchsten Mineralanteil mit bis zu 2600 g/l und ist von den Wassertypen Na-Cl-HCO_3, sowie Na-Ca-Cl geprägt. Generell nimmt die Mineralisation vom Nördlichen Beckenrand zu Südwestlichen Becken hin, zu.

Im Rahmen den „PETher"-Projektes wurden drei Standorte im Molassebecken untersucht, die das Oberjura-Aquifer abgedecken. So konnte die Gruppe II mit dem Wassertyp Na-Ca-HCO_3-Cl, sowie die Gruppe III mit dem Na-HCO_3-Cl-Typ analysiert werden. Die Abbildungen 1-3 geben die in-situ gemessen Thermalwasserdaten hinsichtlich Wärmekapazität, Viskosität und Dichte für die beiden Wassertypen des Molassebeckens wieder. Der Gesamtsalzgehalt (TDS) ist jeweils angegeben. Zum Vergleich sind die Daten reinen Wassers bei p=2,1 MPa in den Diagrammen eingetragen, [13]. Die Messgenauigkeit liegt bei 1% (bezogen auf die Tabellenwerte des reinen Wassers). Wie aus den Abbildungen 1-3 ersichtlich, weichen die Stoffdaten der gemessenen Thermalwässer des Molassebeckens unwesentlich von den Stoffdaten reinen Wassers ab, was deren niedrige Salinität vermuten lässt.

Oberrheingraben

Die für die geothermale Nutzung infrage kommenden Nutzhorizonte des Oberrheingrabens sind im Wesentlichen die Formationen des Oberen Muschelkalks, des Buntsandsteins sowie im südlichen Grabenabschnitt des Hautrogensteins (Mitteljura). Ebenso kommen die Formation des Rotliegenden, die oberste Zone des kristallinen Grundgebirges und die im nördlichen Grabenabschnitt liegenden kiesigen Sande des Tertiärs in Frage. Die Wässer des Hauptrogensteins sind bis in Tiefen von 500 m zu finden und gehören dem Wassertyp Ca-(Mg)-HCO_3 mit geringen Salinitäten von 1 g/kg an. Lediglich lokal, können diese erheblich erhöht sein. Im Oberen Muschelkalk, der sich bis in Tiefen von 1200 m erstreckt, lassen sich zwei unterschiedliche Wassertypen identifizieren. Bei geringeren Tiefen dominiert der Typ Ca-Na-SO_4-HCO_3 mit vergleichsweise niedriger Mineralisation, während die tieferen Lagen den Wassertyp Na-HCO_3-SO_4-Cl aufweisen. Die Mineralisation ist generell gering bei bis zu 20 g/l, wobei lokal auch höhere Werte vorliegen können. Die Buntsandstein Formation kann in Tiefen zwischen 2000 – 3000 m angetroffen werden. In den oberen Schichten dominieren hier Kalziumhydrogencarbonat und Sulfate, während in den unteren Schichten vermehrt Natriumchlorid vorliegt. Der Salzgehalt ist tendenziell hoch und erreicht Werte von bis zu 130 g/l. Salinitäten in gleicher Größenordnung weisen die Wässer des Kristallin auf, die von Natriumchlorid und Kalziumchlorid geprägt sind, [12, 14] und [15]. Aufgrund des kluftreichen Untergrunds tritt im Oberrheingraben eine Zumischung von Tiefenwässern in höher gelegene Schichten auf, so dass häufig Mischwasser mit erhöhtem Natriumchloridanteil zu finden ist.

Um die unterschiedlichen Wassertypen des Oberrheingrabens zu untersuchen, wurden vier Standorte ausgewählt. Dabei konnte der Muschelkalkaquifer mit dem Wassertyp Ca-Na-SO_4-HCO_3, sowie die Natriumchloridreichen Wässer des Kristallin untersucht werden.

Die experimentell ermittelten Daten sind in den Abbildungen 1-3 dargestellt. Anhand dieser Diagramme wird ersichtlich, dass niedrige Salzgehalte die physikalischen Stoffwerte kaum beeinflussen. Die Abweichungen gegenüber den Wasserwerten liegen innerhalb der Messgenauigkeit. Größere Abweichungen zu den Wasserdaten von mehr als 1% ergeben sich erst bei Salzgehalten von mehr als 10 g/l. Dies betrifft alle drei gemessenen Stoffdaten.

Norddeutsches Becken

Die Thermalwässer des Norddeutschen Beckens weisen sehr hohe Salinitäten auf. Diese sind primär aus salinaren Porenwässern

(Brack- oder Meerwässer) sowie aus angereicherten, evaporitischen Wässer entstanden, [14]. Von der Unterkreide bis zum Rotliegenden und Karbon/Devon beträgt die Salinität mehr 100 g/l. Dabei weisen die Wässer des Schilfsandsteins, des Zechsteins und des Rotliegenden die höchsten Werte auf, die 300 g/l und mehr betragen können. Von der Unterkreide bis zum Rhät liegt der Wassertyp Na-Cl vor.

In tieferen Lagen jenseits des Rhät sind die Wässer mit Kalzium angereichert, insbesondere, beim Vorhandensein von Feldspäten, so dass dort der Wassertyp Na-Ca-Cl bzw. Ca-Na-Cl dominiert. Der Anteil an Sulfaten und Karbonaten ist gering, höhere Magnesiumanteile sind im Rhät und im Malm zu finden. Die Standorte im Norddeutschen Becken, die für die in-situ Messungen ausgewählt wurden, haben eine Teufe, die bis in die Rhät Formation reicht. Geothermiestandorte, die das Thermalwasser aus dem Rotliegenden nutzen, standen nicht zur Verfügung. Die in-situ gemessenen Stoffwerte des Norddeutschen Beckens sind ebenfalls in den Abbildungen 1-3 eingetragen.

Wie aus den Diagrammen ersichtlich, nimmt die Wärmekapazität mit zunehmender Salinität ab, während die kinematische Viskosität und die Dichte ansteigen. Dabei werden die Stoffwerte primär vom Gesamtsalzgehalt (TDS) beeinflusst. Der Einfluss der Ionensorte auf die Stoffwerteänderung ist im Rahmen der in-situ durchgeführten Messungen nicht ermittelbar, da die höher konzentrierten Wässer alle eine starke Dominanz von Natriumchlorid aufweisen und sich daher in der Zusammensetzung ähneln, so dass der Einfluss weiterer Ionen im Rahmen der Messgenauigkeit nicht zutage tritt. Nach umfangreicher Datenanalyse wurden von den Projektpartnern typische Thermalwasserzusammensetzungen identifiziert, welche als Grundlage zur Herstellung künstlicher Wässer dienten. Diese wurden im Labor hinsichtlich der physikalischen Stoffeigenschaften untersucht. Mit Hilfe der in-situ gemessenen Daten und den Werten der künstlichen Wässer konnten schließlich vom Projektpartner GeoT Kennkurven erstellt werden.

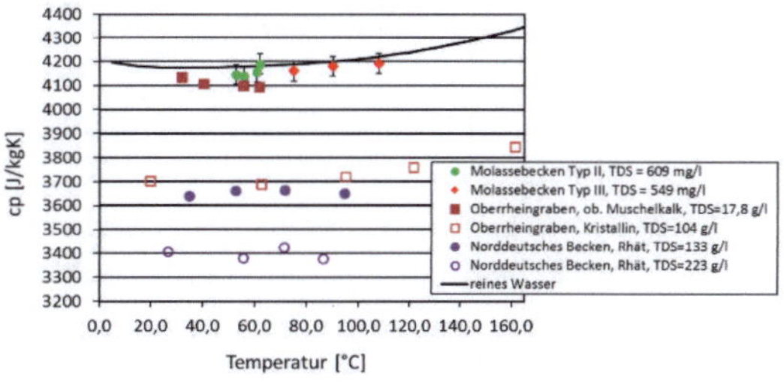

Abb. 1: In-situ gemessene spezifische, isobare Wärmekapazität verschiedener Geothermiestandorte.

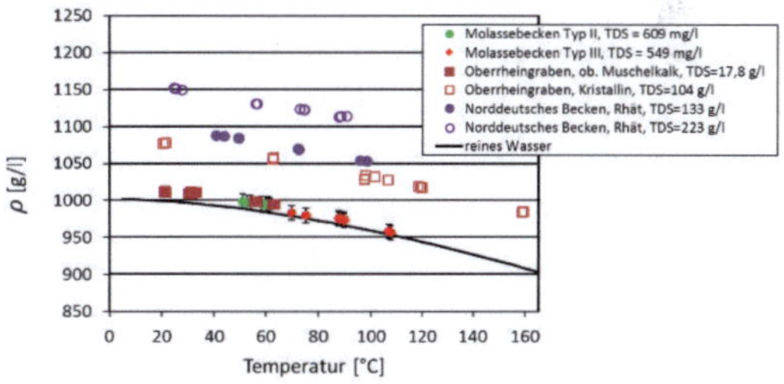

Abb. 2: In-situ gemessene Fluiddichtdichte verschiedener Geothermiestandorte

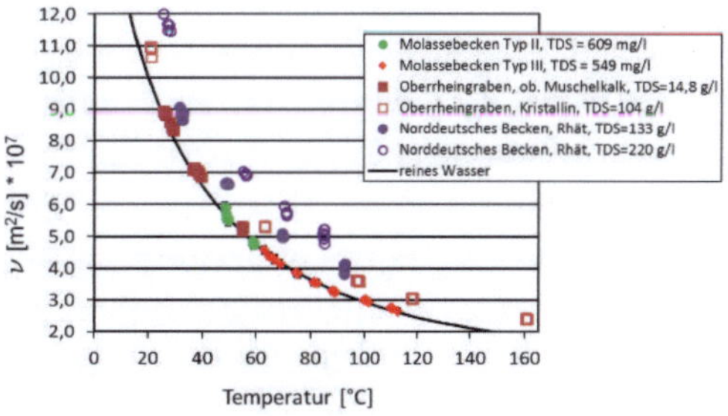

Abb. 3: In-situ gemessene spezifische, isobare Wärmekapazität und Fluiddichtdichte verschiedener Geothermiestandorte.

Erstellung von Kennkurven

Mit den Messdaten der synthetischen Wässer und der vor Ort gemessenen Thermalwässer wurden Kennkurven modelliert. Hierbei kamen zwei Interpolationsschritte zum Einsatz. Im ersten Schritt wurden die Werte einer zusammengehörigen Messreihe mittels eines Polynoms (Dichte: 2. Grades; kinetische Viskosität: 4. Grades; spezifische Wärmekapazität: 2. Grades) über den gesamten Temperaturbereich zu Modellierungsstützkurven verbunden. Im zweiten Schritt wurden durch lineare Interpolationen über alle Stützkurven die übrigen Kurven für die nicht gemessenen Salinitäten in den Bereichen zwischen den Stützkurven erstellt. In den Abbildungen 4-6 sind die Kennkurven basierend auf dem Gesamtsalzgehalt in Salinitätsschritten von 25 g/l dargestellt. Die detaillierte Beschreibung der Kennkurvenerstellung, sowie der experimentellen Arbeiten ist in [16] und [10] zu finden. Mit Hilfe der Kennkurven und der chemischen Analyse der Thermalwässer lassen sich die für die energetische Kraftwerksauslegung notwendigen thermodynamischen Stoffgrößen abschätzen.

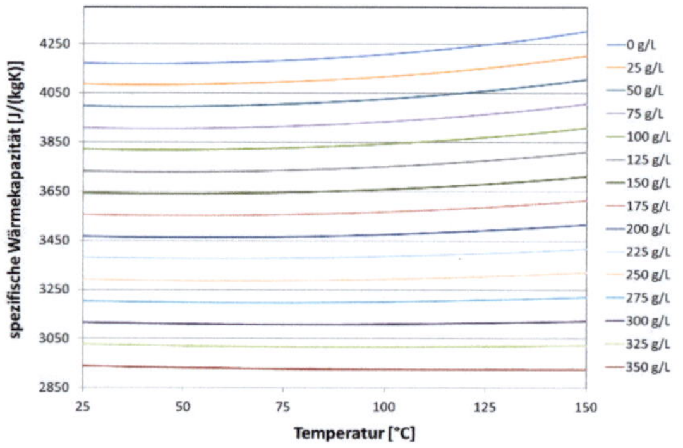

Abb. 4: Abhängigkeit der spezifischen, isobaren Wärmekapazität von der Salinität der Thermalwässer, basierend auf in-situ Messungen und Modellwasseruntersuchungen.

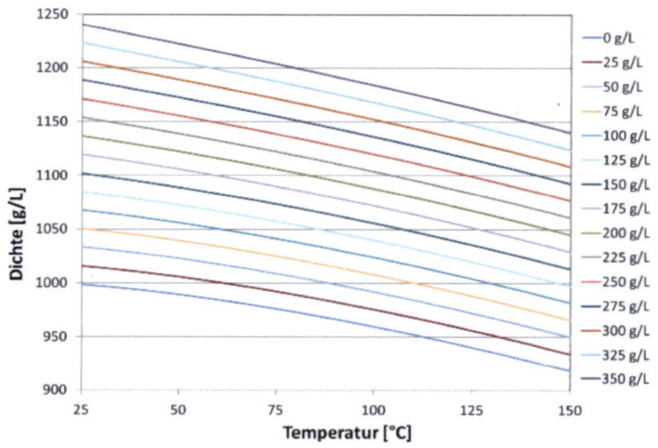

Abb. 5: Abhängigkeit der Dichte von der Salinität der Thermalwässer, basierend auf in-situ Messungen und Modellwasseruntersuchungen

Zusammenfassung

Im Rahmen der vorliegenden Arbeit konnten die thermodynamischen Stoffdaten, wie spezifische, isobare Wärmekapazität, kinematische Viskosität und die Dichte unter Anlagenbedingungen bestimmt werden. Dazu wurden zehn repräsentative Geothermiestandorte ausgewählt und beprobt. Die in-situ Messungen wurden durch Messreihen mit Modellwässern ergänzt und Kennkurven für die verschiedenen Aquifere erstellt. Die erzielten Ergebnisse leisten einen Beitrag zur Verbesserung der Datenlage bei und erlauben es im Rahmen künftiger Geothermieprojekte, die energetische Optimierung der Anlagen voranzubringen.

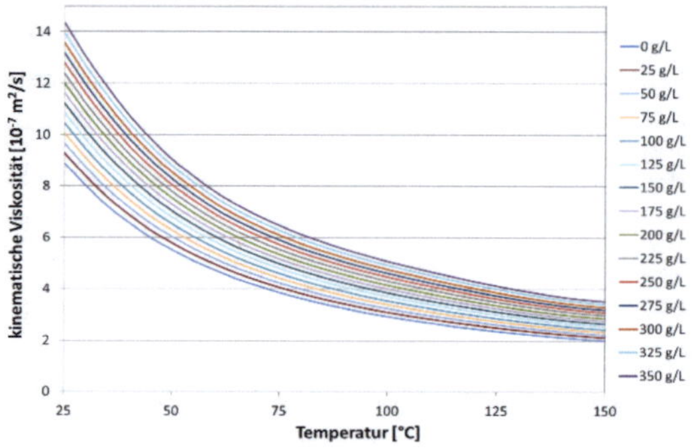

Abb. 6: Abhängigkeit der kinematischen Viskosität von der Salinität der Thermalwässer, basierend auf in-situ Messungen und Modellwasseruntersuchungen.

Partner:

Wir bedanken uns bei den Projektpartnern Geothermal Engineering (GeoT), Geothermie Neubrandenburg (GTN) und Global Engineering & Consulting-Company GmbH (gec-co) für die konstruktive und erfolgreiche Zusammenarbeit.

Literatur

[1] Archer, D. G., Thermodynamic Properties of the NaCl+H_2O System. II. Thermodynamic Properties of NaCl(aq), NaCl*2H_2O(cr), and Phase Equilibria. *Journal of Physical and Chemical Reference Data*, vol. 21, no. 4, pp. 793-829, 1992.

[2] Chou, J. C. S. and Rowe, A. M., Enthalpies of aqueous sodium chloride solutions from 32 to 350°F. *Desalination*, vol. 6, no. 1, pp. 105-115, 1969.

[3] Correia, R. J., Kestin, J. and Khalifa, H. E., Measurement and Calculation of the Viscosity of Mixed Aqueous Solutions of NaCl and KCl in the Temperature Range 25°C-150°C and the Pressure Range 0–30 MPa. *Berichte der Bunsengesellschaft für physikalische Chemie*, vol. 83, no. 1, pp. 20-24, 1979.

[4] Gates, J. A., Tillett, D. M., White, D. E. and Wood, R. H., Apparent molar heat capacities of aqueous NaCl solutions from 0.05 to 3.0 $mol{\cdot}kg^{-1}$, 350 to 600 K, and 2 to 18 MPa. *The Journal of Chemical Thermodynamics*, vol. 19, no. 2, pp. 131-146, 1987.

[5] Jamieson, D. T., Tudhope, J. S., Morris, R. and Cartwright, G., Physical properties of sea water solutions: heat capacity. *Desalination and Water Treatment*, vol. 7, no. 1, pp. 23-30, 1969.

[6] Kestin, J., Khalifa, H. E. and Correia, R. J., Tables of the dynamic and kinematic viscosity of aqueous NaCl solutions in the temperature range 20-150 °C and the pressure range 0.1-35 MPa. *Journal of Physical and Chemical Reference Data*, vol. 10, no. 1, pp. 71-88, 1981.

[7] Mao, S. and Duan, Z., The Viscosity of Aqueous Alkali-Chloride Solutions up to 623 K, 1,000 bar, and High Ionic Strength. *International Journal of Thermophysics*, vol. 30, no. 5, pp. 1510-1523, 2009.

[8] Nayar, K. G., Sharqawy, M. H., Banchik, L. D. and Lienhard V, J. H., Thermophysical properties of seawater: A review and new correlations that include pressure dependence. *Desalination*, vol. 390, no. pp. 1-24, 2016.

[9] Sharqawy, M. H., New correlations for seawater and pure water thermal conductivity at different temperatures and salinities. *Desalination*, vol. 313, no. 0, pp. 97-104, 2013.

[10] Schröder, E., Thomauske, K., Kuhn, D., Thorwart, K., Nowak, K. and Markus, W., BMWi-Verbundprojekt PEther "Physikalische Eigenschaften von Thermalwasser unter in-situ-Bedingungen", Teilvorhaben 0325761A: "Experimentelle Bestimmung der physikalischen Eigenschaften von Thermalwasser in situ und im Labor": Abschlussbericht: Laufzeit des Vorhabens: 01.03.2015 bis 31.12.2017, Karlsruhe, 2018.

[11] Schröder, E., Thomauske, K., Schmalzbauer, J. and Herberger, S., Measuring Techniques for in Situ Measurements of Thermodynamic Properties of Geothermal Water, *World Geothermal Congress*, Melbourne, Australia, 2015.

[12] Birner, J., Hydrogeologisches Modell des Malmaquifers im Süddeutschen Molassebecken, -thesis, Freie Universität Berlin, Berlin, 2013.

[13] NIST Reference Fluid Thermodynamic and Transport Properties—REFPROP, 9, *Physical and Chemical Properties Division, National Institute of Standards and Technology*, Boulder, Colorado 80305 U.S., 2010.

[14] Wolfgramm, M., Thorwart, K., Rauppach, K. and Brandes, J., Zusammensetzung, Herkunft und Genese geothermaler Tiefengrundwässer im Norddeutschen Becken (NDB) und deren Relevanz für die geothermische Nutzung. *Zeit. Geol. Wiss.*, vol. 39, no. 3 - 4, pp. 173 - 193, 2011.

[15] Stober, I., *Die Wasserführung des kristallinen Grundgebirges*, Ferdinand Enke Verlag, Stuttgart, 1995.

[16] Jodocy, M. and Kraml, M., BMWi-Verbundprojekt PETher: physikalische Eigenschaften von Thermalwasser unter In-situ-Bedingungen: Teilvorhaben B: Modellierung von Kennkurven zur Beschreibung physikochemischer Eigenschaften von Thermalwasser: Abschlussbericht Projektphasen I & V: Datenrecherche und Typisierung der Thermalwässer im Oberrheingraben und Süddeutschen Molassebecken und Modellierung mit Integration und Synthese: Laufzeit des Vorhabens: 01.03.2015-31.12.2017, Karlsruhe, 2018.

Group: Framatome Professional School

CFD validation of fuel assembly flow

SESAME Deliverable 854935 D2.18

Abdalla Batta, Andreas Class

Introduction

The development and validation of advanced numerical approaches for liquid metal fast reactors is one of the mean objective of SESAME project. The work presented here is a part of work package 2 (WP2) where RANs Models is validated with DNS results generated within SESAME project. The considered case represents flow of liquid metal in infinite heated fuel pin bundle simulator (FPBS). The DNS results (three cases) were presented in [1,2]. The considered geometrical parameters are similar to those used in the fuel pin simulator test section built in the NACIA-UP facility located at ENEA. See [3,4] for more details about the experiment and test cases. Figure 1 shows a sketch of the fuel pin bundle simulator and a picture of the teste section. This fuel pin bundle configuration is relevant for the ALFRED's core thermal hydraulic design. ALFRED is a flexible fast spectrum research reactor (300 MWth).

The main geometrical dimensions to be considered for a thermal-hydraulic assessment of the FA are:

- The rod diameter *D*=10 mm;
- The pitch to diameter ratio *P/D* =1.4,
- The distance between the last rank of pins and the internal wall of the wrap δw= 1.75 mm;
- The regular lattice is triangular/hexagonal staggered;
- The internal hexagonal key of the wrapper is *H*=62 mm.

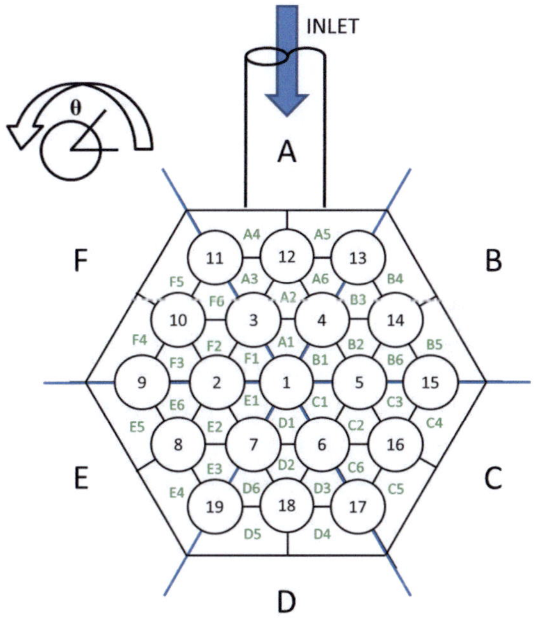

Fig. 1 top view of the fuel pin bundle simulator for the NACIE facility, schematic view (top), actual picture (bottom), [2].

Table 1 summarize the considered cases for DNS study and figure 2 illustrate the considered periodical element. Data are taken from [1,2]. The DNS results were obtained for

Pr=0.31 computed at a reference temperature T_{ref} =220°C. This temperature is consided as the average fluid temperature in the computed domain. The fluid properties are according to [5]. For case 1 and case 2 of table 1 constant wall heat flux of 65.8 kW/m². For case 3 the heat flux q" is 131.6 kW/m². The gravity is only consider in case 2 where in case 1 and 3 the gravity effect is not considered. The red marked region in figure 2 is considered for the DNS computations. The computational domain size considered is $1.2D_h \times 2.1D_h \times 8\pi D_h$, [2]. The hydraulic diameter (D_h) of the subchannel triangular infinite bundle is given by the following equation

$$D_h = \left[\frac{2\sqrt{3}}{\pi} * \left(\frac{P}{D}\right)^2 - 1\right] D \qquad \text{(Eq. 1)}$$

Using the given geometrical parameters of P/D=1.4 the subchannel hydraulic diameter for the present case is 1.16D. The DNS results are compared to Kirillov et. el correlation [6] in Table 1. Good agreement can be seen for Nusselt number (Nu). The correlation is simplified as follows:

$$Nu = 7.55X - 20/X^{13} + (0.041/X^2)\, Pe^{0.56+0.19X}$$
(Eq. 2)

Where X=P/D and Pe=Re.Pr. The accuracy of this correlation is estimate by [6] to be 12-15%. This correlation was selected for the comparison, because it represents near average values of many available correlation in literature according to Manservisi [7] comparison.

Table 1. DNS considered cases and resulted Nusselt number.

	Case 1	Case 2	Case3
Re_T	550	550	1100
Ri	0	0.25	0
Pr	0.031	0.031	0.031
Re_b	8290	8660	16260
Nu DNS	12.82	12.65	14.79
Nu Kirillov [6]	12.37	12.44	13.89

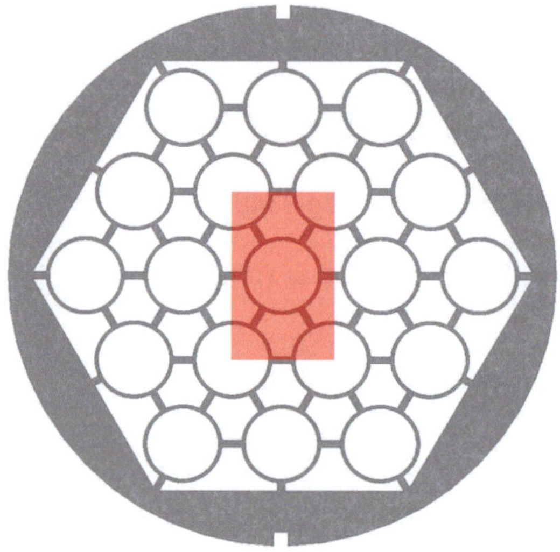

(a)

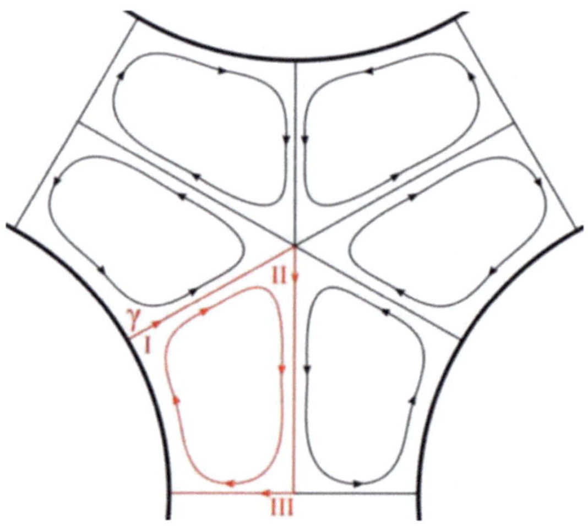

(b)

Fig. 2 Periodic module of the NACIE bundle: (a) cross flow layout with periodic rectangular module highlighted, covering four subchannels; (b) unit flow cells, with indication of the curvilinear abscissa γ, defined along its border [2].

In the present work, LBE physical properties, which are all temperature-dependent, are evaluated using empirical correlations suggested in (OECD/NEA Handbook, (2015)). Formulas for density, specific heat, dynamic viscosity, conductivity are reported in Table 2.

Table 2 Physical properties of LBE as a function of temperature (T in Kelvin).

Property	Symbol	Correlation	Maximum Uncertainty	Standard deviation
Density	$\rho(T)$	$11065 - 1.293 \cdot T$	$\leq 0.8\%$	0.58%
Heat capacity	$c_p(T)$	$164.8 - 3.94 \cdot 10^{-2} \cdot T + 1.25 \cdot 10^{-5} \cdot T^2 - 4.56 \cdot 10^{-5} \cdot T^{-2}$	$\leq 5.0\%$	2.4%
Dynamic viscosity	$\mu(T)$	$4.94 \cdot 10^{-4} \exp(754.1/T)$	$\leq 6.0\% - 8.0\%$	7.2%
Thermal conductivity	$k(T)$	$3.284 + 1.67 \cdot 10^{-2} \cdot T - 2.305 \cdot 10^{-6} T^2$	$\leq 10.0\%$ -15.0%	6.2%

In the next section numerical RANs model, computational domain and the numerical Nusselt number results will be presented and compared to the reference DNS results presented above.

Numerical model and results

The selected periodical element shown in figure 2-a is considered for the RANS simulation. An axial length equivalent to the heated length of NACIA experiment is considered, heated length L_h is 0.6 m. Fig. 3 shows the heated walls of the computational domain with the y+ contours for case 1. As it can be seen y+ is selected to be less than 0.5 so that when the higher velocity is considered for case 3 the y+ remains under 1 to satisfy the model needs for low y+ values. Near 5 million cells were used

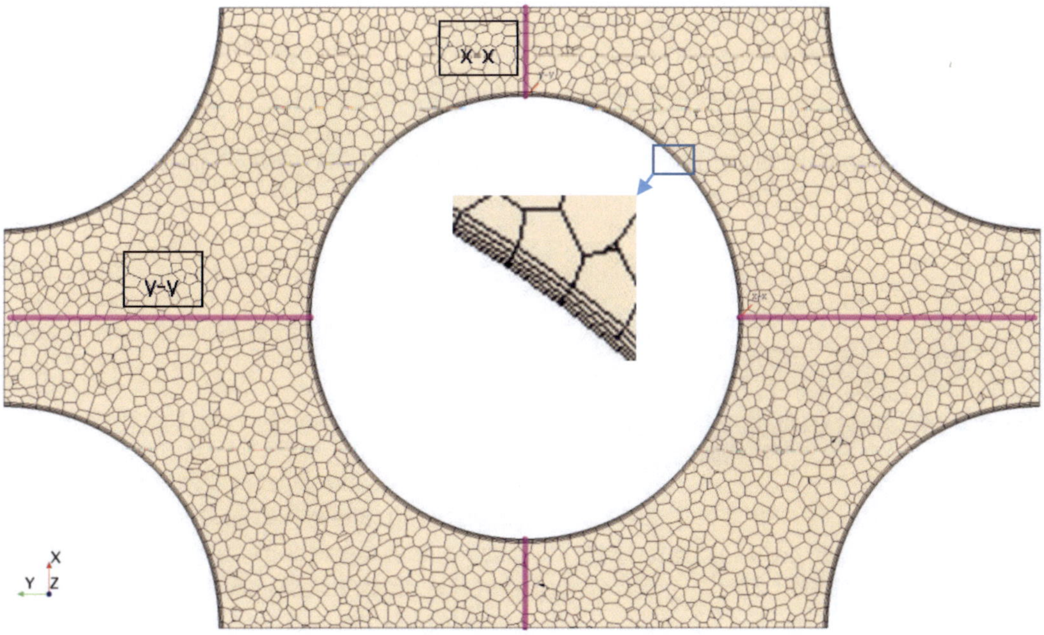

Fig. 4. Mesh and lines through centre plane.

to mesh the domain. Figure 4 shows a cross section of the mesh at the middle of the computational domain. It shows also two lines which will be used in the post processing.

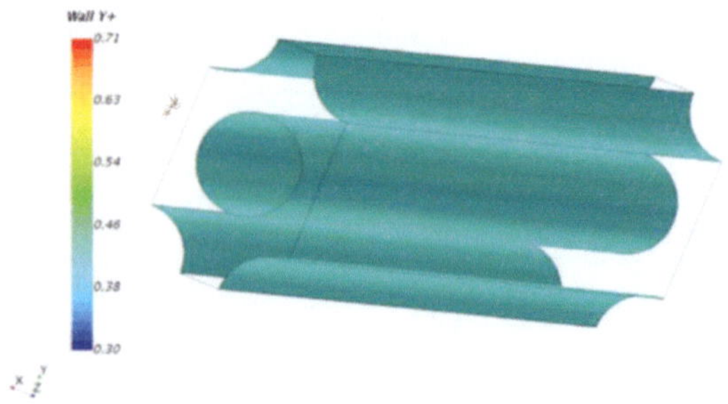

Fig. 3. Y+ values for case 1 and computational domain.

CCM+ is used for the simulation of the current study. The standard k-ε low-Re turbulent model with all y+ wall treatment for steady flow is applied. Fig. 5 shows the velocity contours at inlet and the outlet of the periodical element considered. It shows the walls of the heated rods (0 velocity) one of the rods is not illustrated for purpose of clearness. Local temperature along two lines passing through the center of the computational domain as illustrated in Fig. 4 are plotted in figure 6. It can be seen that, with the considered fine mesh, approximately linear temperature distribution adjacent to the walls can be seen. Temperature contours for case 1 is given in Fig. 7, it shows the temperature contours at inlet, outlet and at rods.

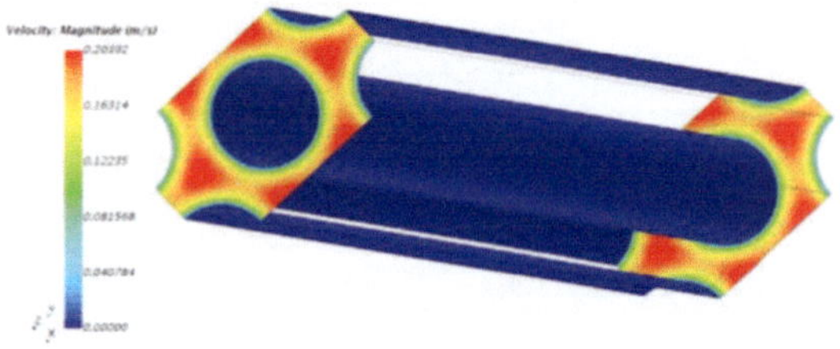

Fig. 5. Velocity contours for case 1. One rod wall is not illustrated for clearness

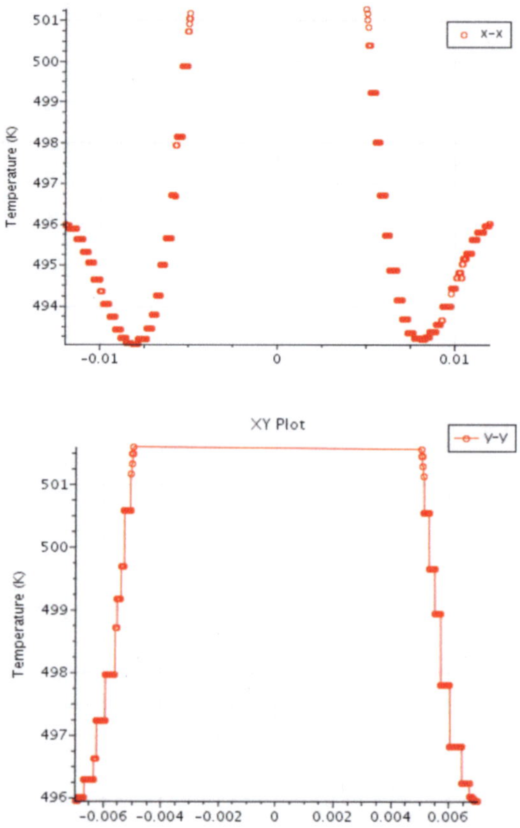

Fig. 6. Temperature along x-x and y-y lines shown in Figure4.

Fig. 7. Temperature contours for case 1, without gravity.

For case 2 and 3 temperature contours are given in Fig. 8 and Fig. 9. Comparing .Fig. 7 for case 1 with figure 8 for case 2, it can be seen that the gravity has very small effect on temperature distribution, so that near similar distribution of temperature as in case 1 was observed. The maximum difference in local temperature was less than 3°C, compare the scales of Fig. 7 and Fig. 8. Considering case 3, which has double mass flow rate and double heat load as case 1, the temperature contours shown in Fig. 9 shows that case 3 has higher wall temperatures than case1, near 8 °C.

Fig. 8. Temperature contours including gravity, case2.

Fig. 9. Temperature contours without gravity, case3.

Comparison of RANs and DNS results

Nusselt number is calculated based on the average wall temperature and average bulk temperature using fluid properties according to table 2 at mean bulk temperature. Table 3 present the numerical results compared to DNS and Kirillov [6] results. The compression of both DNS results and the RANS results give the same qualitative agreement with [6]. The comparison is plotted in figure 10.

Table 3. Nusselt number values compared to DNS and Kirillov[6].

Pe	Ri	DNS	RANs	[6]
257	0	12.82	12.68	12.37
268	0.25	12.65	13.00	12.44
504	0	14.79	14.15	13.89

Conclusion

The numerical study conducted here has compared the heat transfer results obtained from standard RANS model to the DNS results and existing correlation of [6]. The comparison of Nusselt number for all compared cases show a very good agreement with the numerical DNS result and with the literature. Considering the effect of gravity by comparing case 1 and case 2, the resulted temperature distribution looks similar for the given difference of Richardson number(Ri=0.25) . Comparison of case 1 to case 3 which has double mass flow and heat load than case 1 has shown that slight increase in the maximum domain temperature of near 8°C.

Considering the computational domain for periodical element, one could use shorter computational domain since the domain length in z direction shows a periodical behaviour. Considering the velocity distribution in the x-y plane it can be seen that a very smaller periodical element like illustrated in red in figure 2-b could be sufficient for the periodical study.

Nomenclature

Symbol	Quantity	SI Unit
D	Pin diameter	m
D_h	Hydraulic diameter	m
L_h	Heated length	m
H	Hexagonal key of the wrapper	m
Nu	Section average Nusselt number	-
P	Pitch of the bundle	m
p, p_m	Pressure, modified pressure fields	Pa
Pr	Prandtl number	-

q''	Wall heat flux	W/m^2
Re_b	Bulk Reynolds number	-
Re_τ	Reynolds tau number	-
Ri	Richardson number	-
T	Temperature	°C
T_{ref}	Reference temperature difference	°C
T_{bulk}	Bulk temperature	°C
u	Dimensionless velocity component	-
x,y,z	Dimensionless spatial coordinates	-
Greek letters		
δ_w	Distance between the last pin and the wrapper	m
k	Thermal conductivity	W/m/K
μ	Dynamic viscosity	Pa s
ρ	Density	kg/m^3

References

[1] Diego, A., MIXED CONVECTION OF LBE IN THE FUEL PIN BUNDLE OF THE NACIE-UP FACILITY: EXPERIMENTS AND SIMULATIONS SESAME workshop, Petten 2019

[2] Diego, A., Reference data for rod bundle, UNIMORE, SESAME deliverable, D2.6

[3] Di Piazza, I., BFPS Experiment, ENEA, SESAME deliverables, D2.7

[4] Di Piazza, I., Experiment set-up BFPS, ENEA, SESAME deliverables, D2.5

[5] OECD and N. E. Agency, 2015. Handbook on Lead-bismuth Eutectic Alloy and Lead Properties , Materials Compatibility , Thermal- hydraulics and Technologies.

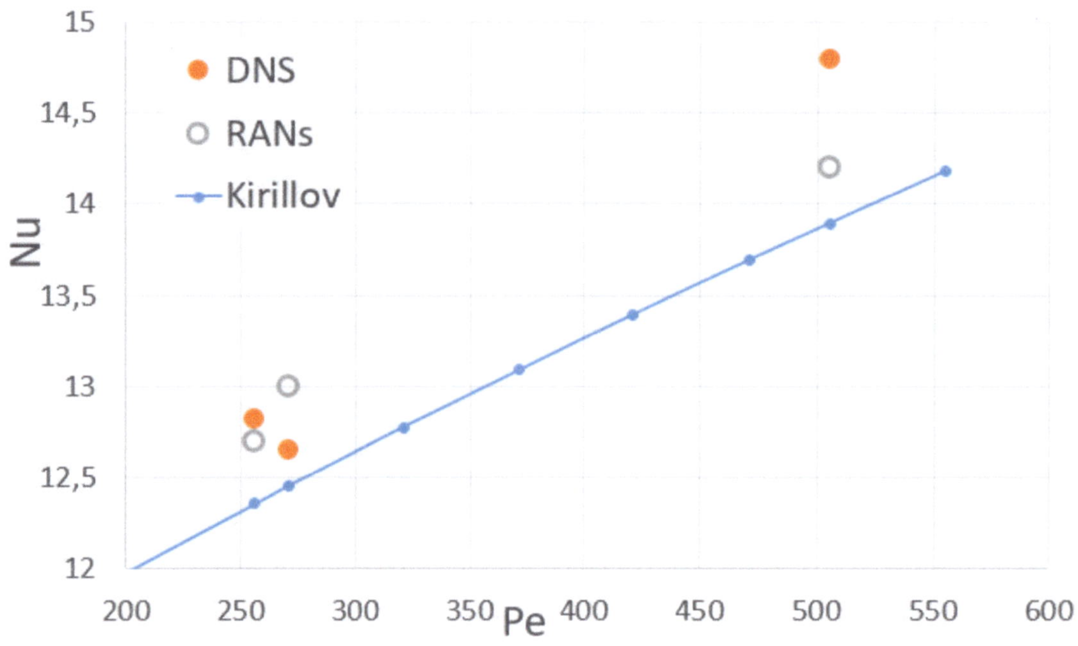

Fig. 10. Nusselt number comparison between RANS results obtained in this study, DNS [1,2] and Kirrillov [6]

[6] Kirillov, P.L. and Ushakov, P.A., 2001. Liquid–metal heat transfer in rod bundles. Thermal Engineering, vol 48, pp.127–133. Marinari, R., Di Piazza, I., Forgione, N., and Magugliani, F., 2017. Pre-test CFD simulations of the NACIE-UP BFPS test section

[7] Manservisi, S. and Menghini, F., Triangular rod bundle simulations of a CFD K-ε-K-Θ -ε-Θ heat transfer turbulence model for heavy liquid metals, Nuclear Engineering and Design 273 (2014) 251–270

List of Publications

Publications 2017

[1] Arlt, T.; Priede, J.; Bühler, L. The effect of finite-conductivity Hartmann walls on the linear stability of Hunt's flow. 2017. Journal of fluid mechanics, 822, 880–891. doi:10.1017/jfm.2017.322

[2] Arlt, T.; Priede, J.; Bühler, L. Influence of thin finite conducting walls on the linear stability of magnetohydrodynamic duct flow. 2017. Magnetohydrodynamics, 53 (1), 35–44

[3] Batta, A.; Class, A. G., Numerical analysis of MYRRHA inter-wrapper flow experiment at KALLA.48. Jahrestagung Kerntechnik / 48th Annual Meeting on Nuclear Technology (AMNT 2017), Berlin, May 16-17, 2017, CD-ROM, INFORUM GmbH, Berlin

[4] Batta, A.; Class, A.; Pacio, J. Numerical analysis of a LBE-cooled blocked 19-pin hexagonal wire wrapped rod bundle experiment carried out at KIT-KALLA within EC-FP7 project MAXSIMA.17th International Topical Meeting on Nuclear Reactor Thermal Hydraulics (NURETH-17), Xi'an, China, September 3-8, 2017. Proceedings on USB-Stick, Paper No.20532

[5] Benz, M.; Schulenberg, T. Statistical Modeling of Stratified Two-Phase Flow. 2017. Journal of Nuclear Engineering and Radiation Science, 3 (2), Art. Nr. 021001. doi:10.1115/1.4035564

[6] Breitung, W.; Halmer, G.; Xiao, J.; Kuznetsov, M.; Jordan, T. Analysis of transient supersonic hydrogen release, dispersion and combustion.7th International Conference on Hydrogen Safety (ICHS-2017), Hamburg, September 11-13, 2017

[7] Bühler, L.; Aiello, G.; Bendotti, S.; Koehly, C.; Mistrangelo, C.; Galabert, J. Development of combined temperature - electric potential sensors. 13th International Symposium on Fusion Nuclear Technology (ISFNT), Kyoto, J, September 25-29, 2017

[8] Bühler, L.; Arlt, T.; Boeck, T.; Braiden, L.; Chowdhury, V.; Krasnov, D.; Mistrangelo, C.; Molokov, S.; Priede, J. Magnetically induced instabilities in duct flows. 2017. IOP conference series / Materials science and engineering, 228, Art. Nr. 012003. doi:10.1088/1757-899X/228/1/012003

[9] Bühler, L.; Arlt, T.; Chowdhury, V.; Mistrangelo, C. Instabilities in boundary layers in MHD duct flows. 7th International Conference on Magneto-Science (ICMS 2017), Reims, F, October 23-27, 2017. Proceedings on USB-Stick, Paper No KN2, S. 2–3

[10] Chen, X. N.; Gabrielli, F.; Rineiski, A.; Schulenberg, T. Possibility studies of a boiling water cooled traveling wave reactor. International Conference on Fast Reactors and Related Fuel Cycles: Next Generation Nuclear Systems for Sustainable Development (FR17), Yekaterinburg, Russia, June 26-29, 2017. Proceedings on USB-Stick, Paper IAEA-CN245–276

[11] Chen, X.-N.; Li, R.; Belloni, F.; Gabrielli, F.; Rineiski, A.; Andriolo, L.; Guo, L.; Castelliti, D.; Schyns, M.; Bubelis, E.; Bandini, G.; Sarotto, M. Safety studies for the MYRRHA critical core with the SIMMER-III code. 2017. Annals of nuclear energy, 110, 1030–1042. doi:10.1016/j.anucene.2017.08.021

[12] Dietrich, B.; Marocco, L.; Wetzel, T. Wärmeübertragung in über den halben Umfang beheizten, mit Flüssigmetall durchströmten Rohren. Jahrestreffen der Process-Net Fachgruppe Wärme- und Stoffübertragung, Bruchsal, 16.-17.Februar 2017

[13] Doolaard, H.; Roelofs, F.; Pacio, J.; Batta, A. Experiment design to assess the inter-wrapper heat transfer in LMFR.

17th International Topical Meeting on Nuclear Reactor Thermal Hydraulics Xi'an, Shaanxi, China, Sept. 3-8, 2017. Proceedings on USB-Stick, Paper No.20279

[14] Fichot, F.; Carenini, L.; Sangiorgi, M.; Hermsmeyer, S.; Miassoedov, A.; Bechta, S.; Zdarek, J.; Guennadou, D. Status of the IVMR project: First steps towards a new methodology to assess in-vessel retention strategy for high-power reactors. 8th European Review Meeting on Severe Accident Research (ERMSAR), Warszawa, PL, May 16-18, 2017. Proceedings on USB-Stick, Paper 503

[15] Fieg, G.; Werle, H.; Huber, F.; Foit, J. J. Spreading on wet substrates in KATS experiments. 8th European Review Meeting on Severe Accident Research (ERMSAR), Warszawa, PL, May 16-18, 2017

[16] Flad, M.; Gabrielli, F.; Gianfelici, S.; Li, R.; Maschek, W.; Matzerath-Boccaccini, C.; Vezzoni, B.; Rineiski, A. Quantitative evaluation of the post disassembly energetics of a hypthetical core disruptive accident in a sodium cooled fast reactor. International Conference on Fast Reactors and Related Fuel Cycles: Next Generation Nuclear Systems for Sustainable Development (FR17), Yekaterinburg, Russia, June 26-29, 2017. Proceedings on USB-Stick, Paper IAEA-CN245–483

[17] Flesch, J.; Niedermeier, K.; Fritsch, A.; Musaeva, D.; Marocco, L.; Uhlig, R.; Baake, E.; Buck, R.; Wetzel, T. Liquid metals for solar power systems. 2017. IOP conference series / Materials science and engineering, 228 (1), Art. Nr.: 012012. doi:10.1088/1757-899X/228/1/012012

[18] Foit, J. J. Concrete decomposition temperature relevant to MCCI process on LCS concrete. 8th European Review Meeting on Severe Accident Research (ERMSAR), Warszawa, PL, May 16-18, 2017. Proceedings on USB-Stick, Paper 202

[19] Fritsch, A.; Uhlig, R.; Marocco, L.; Frantz, C.; Flesch, R.; Hoffschmidt, B. A comparison between transient CFD and FEM simulations of solar central receiver tubes using molten salt and liquid metals. 2017. Solar energy, 155, 259–266. doi:10.1016/j.solener.2017.06.022

[20] Gabrielli, F.; Flad, M.; Gianfelici, S.; Li, R.; Kriventsev, V.; Maschek, W.; Boccaccini, C. M.; Vezzoni, B.; Rineiski, A. Probabilistic Evaluation of the Post Disassembly Energetics of a Hypothetical Core Disruptive Accident in a Sodium-Cooled SMR by using a Phenomenological Relationship Diagram. 2017. Energy procedia, 131, 222–229. doi:10.1016/j.egypro.2017.09.471

[21] Gaus-Liu, X. Experimental database on heat transfer correlations for homogeneous corium pool during TH steady state. IVMR H-2020 Project Workshop, Alkmaar, NL, June 20-23, 2017

[22] Gaus-Liu, X.; Miassoedov, A. Influence of Boundary Conditions, Vessel Geometry, and Simulant Materials on the Heat Transfer of Volumetrically Heated Melt in a Light Water Reactor Lower Head. 2017. Journal of Nuclear Engineering and Radiation Science, 3 (3), 031003. doi:10.1115/1.4035853

[23] Gaus-Liu, X.; Miassoedov, A.; Journeau, C.; Liao, Y.; Cassiaut-Louis, N. The progress of ALISA project: Access to large infrastructures for severe accidents in Europe and in China. 8th European Review Meeting on Severe Accident Research (ERMSAR), Warszawa, PL, May 16-18, 2017. Proceedings on USB-Stick, Paper 613

[24] Gaus-Liu, X.; Miassoedov, A.; Pandazis, P.; Hollands, T. Investigation of molten corium behavior in lower head under external subcooling and biling conditions within LIVE experimental program and ATHLET-CD simulation. 17th International Topical Meeting on Nuclear Reactor Thermal Hydraulics Xi'an, Shaanxi, China, Sept. 3-8, 2017. Proceedings on USB-Stick, Paper No.20755

[25] Geißler, T. G. Methanpyrolyse in einem Flüssigmetall-Blasensäulenreaktor. Dissertation. 2017. Verlag Dr. Hut, München

[26] Guo, L.; Morita, K.; Tobita, Y. Numerical simulations on self-leveling behaviors with cylindrical debris bed. 2017. Nuclear engineering and design, 315, 61–68. doi:10.1016/j.nucengdes.2017.02.024

[27] Haste, T.; Barrachin, M.; Fichot, F.; Repetto, G.; Miassoedov, A.; Steinbrück, M.; Buck, M.; Bechta, S.; Beuzet, E.; Torkham, M.; Hollands, T.; Journeau, C.; Pontillon, Y.; Lind, T.; Hozer, Z.; Horvath, G. L.; Bottomley, P. D. W., CoreSOAR core degradation SOAR update: Status May 2017. 8th European Review Meeting on Severe Accident Research (ERMSAR), Warszawa, PL, May 16-18, 2017. Proceedings on USB-Stick, Paper 502

[28] Heinze, D.; Schulenberg, T.; Behnke, L., A physically based, one-dimensional three-fluid model for direct contact condensation of steam jets in flowing water. 2017. International journal of heat and mass transfer, 106, 1041–1050. doi:10.1016/j.ijheatmasstransfer.2016.10.076

[29] Heinzel, A.; Hering, W.; Konys, J.; Marocco, L.; Litfin, K.; Müller, G.; Pacio, J.; Schroer, C.; Stieglitz, R.; Stoppel, L.; Weisenburger, A.; Wetzel, T. Liquid Metals as Efficient High-Temperature Heat-Transport Fluids. 2017. Energy technology, 5 (7), 1026–1036. doi:10.1002/ente.201600721

[30] Heitkam, S.; Tschisgale, S.; Krull, B.; Wetzel, T.; Baake, E.; Fröhlich, J. Using Lorentz forces to control the distribution of bubbles in a vertical tube filled with liquid metal. 2017. IOP conference series / Materials science and engineering, Art. Nr.: 012008. doi:10.1088/1757-899X/228/1/012008

[31] Holler, T.; Kljenak, I.; Hu, P.; Zhai, S.; Komen, M. J.; Kuznetsov, M. CFD-based blind simulation of medium-scale hydrogen deflagration experiment. A New Paradigm in Nuclear Power Safety : Proceedings of the International Congress on Advances in Nuclear Power Plants, ICAPP 2017, Japan, Fukui and Kyoto, 24th - 28th April 2017, Atomic Energy Society of Japan, Tokyo

[32] Hou, B.; Yu, J.; Xu, Z.; Jiang, G.; Zou, Z. Methodological Study on Prediction of Detonation Cell Width with Gaussian Process Regression. 2017. Hedongli-gongcheng, 38 (2), 72–77. doi:10.13832/j.jnpe.2017.02.0072

[33] Jordan, T.; Veser, A., KIT Busshuttle-Service auf Basis von Wasserstoff-Brennstoffzellen-Bussen und einer Wasserstofftankstelle: Statusbericht 29. Juli 2017. 2017. Karlsruhe

[34] Journeau, C.; Bouyer, V.; Casslaut-Louis, N.; Fouquart, P.; Piluso, P.; Ducros, G.; Gossé, S.; Guéneau, C.; Quaini, A.; Fluhrer, B.; Miassoedov, A.; Stuckert, J.; Steinbrueck, M.; Bechta, S.; Kudinov, P.; Ma, W. M.; Sehgal, B. R.; Hozer, Z.; Guba, A.; Manara, D.; Bottomley, D. P.; Fischer, M.; Langrock, G.; Schmidt, H.; Kiselova, M.; Zdarek, J. Safest roadmap for corium experimental research in Europe. 2017. Journal of risk and uncertainty, 4 (3), Art.Nr. 030901. doi:10.1115/1.4037878

[35] Kaiser, F.; Wetzel, T.; Gabriel, S.; Schulenberg, T.; Albrecht, G.; Heiler, W.; Miassoedov, A., Optical void measurement method for stratified wavy two phase flows. 9th World Conference on Experimental Heat Transfer, Fluid Mechanics and Thermodynamics, Foz do Iguaçu, BR, June 11-15, 2017

[36] Kennedy, G.; Mirelli, F.; Lim, J.; Pacio, J., Operational aspects of experimental liquid metal facilities.Thermohydraulics and Chemistry of Liquid Metal Cooled Reactors, April 10-14, 2017. Ed.: F. Roelofs, Von Karman Inst. for Fluid Dynamics, Rhode Saint Genèse

[37] Koehly, C.; Bühler, L., Fabrication Issues of Sandwich-Like Flow Inserts for Circular Pipes. 2017. Fusion science and technology, 72 (4), 660–666. doi:10.1080/15361055.2017.1350477

[38] Krasnov, D.; Boeck, T.; Bühler, L. Simulations of turbulent magnetohydrodynamic duct flows with high-speed shear layers. 16th European Turbulence Conference, Stockholm, S, August 21-24, 2017

[39] Krasnov, D.; Braiden, L.; Bühler, L.; Molokov, S.; Zikanov, O.; Boeck, T., Properties of turbulent and transitional electromagnetic boundary layers in channels with a uniform magnetic field. Jahrestagung der Gesellschaft für Angewandte Mathematik und Mechanik (GAMM 2017), Weimar, 6.-10.März 2017

[40] Kuhn, D., Evaluierung von Ausfällungsinhibitoren für das Molassenbecken 'EvA-M'. Geothermie in der süddeutschen Molasse : Wissenstransfer der Geothermie-Allianz Bayern, München, 26.Januar 2017

[41] Kuznetsov, M.; Xiao, J.; Travis, J. Effect of steam condensation on pressure and temperature under hydrogen jet fire in a vented enclosure. A New Paradigm in Nuclear Power Safety : Proceedings of the International Congress on Advances in Nuclear Power Plants, ICAPP 2017, Japan, Fukui and Kyoto, 24th - 28th April 2017, Atomic Energy Society of Japan, Tokyo

[42] Li, R.; Chen, X.-N.; Andriolo, L.; Rineiski, A., 3D numerical study of LBE-cooled fuel assembly in MYRRHA using SIMMER-IV code. 2017. Annals of nuclear energy, 104, 42–52. doi:10.1016/j.anucene.2017.02.009

[43] Li, Y.; Zhang, H.; Xiao, J., LES simulation of buoyant jet from unintentioned hydrogen release with GASFLOW-MPI. 7th International Conference on Hydrogen Safety (ICHS-2017), Hamburg, September 11-13, 2017

[44] Madokoro, H.; Kretzschmar, F.; Miassoedov, A., Modeling of RPV lower head under core melt severe accident condition using OpenFOAM. A New Paradigm in Nuclear Power Safety : Proceedings of the International Congress on Advances in Nuclear Power Plants, ICAPP 2017, Japan, Fukui and Kyoto, 24th - 28th April 2017, Art.Nr. 17134, Atomic Energy Society of Japan, Tokyo

[45] Marchetti, M., Neutronics Methods for Transient and Safety Analysis of Fast Reactors (KIT Scientific Reports ; 7728). Dissertation. 2017. KIT Scientific Publishing, Karlsruhe. doi:10.5445/KSP/1000063691

[46] Margulis, M.; Blaise, P.; Gabrielli, F.; Gruel, A.; Mellier, F.; Gilad, E., The path for innovative severe accident neutronics studies in ZPRs. Part I.1 - Analysis of SNEAK-12A experiments for core disruption in LMFBRs. 2017. Progress in nuclear energy, 94, 106–125. doi:10.1016/j.pnucene.2016.11.001

[47] Marocco, L.; Alberti di Valmontana, A.; Wetzel, T., Numerical investigation of turbulent aided mixed convection of liquid metal flow through a concentric annulus. 2017. International journal of heat and mass transfer, 105, 479–494. doi:10.1016/j.ijheatmasstransfer.2016.09.107

[48] Massone, M.; Gabrielli, F.; Rineiski, A. SIMMER extension for multigroup energy structure search using genetic algorithm with different fitness functions. 2017. Nuclear engineering and technology, 49 (6), 1250–1258. doi:10.1016/j.net.2017.07.012

[49] Massone, M.; Gabrielli, F.; Rineiski, A. SIMMER extension for multigroup energy structure search using genetic algorithm. International Conference on Mathematics and Computational Methods Applied to Nuclear Science and Engineering (M&C 2017), Jeju, Korea, April 16-20, 2017. Proceedings on USB-Stick, P273S06–04MassoneMVE

[50] Massone, M.; Gabrielli, F.; Rineiski, A. A genetic algorithm for multigroup energy structure search. 2017. Annals of nuclear energy, 105, 369–387. doi:10.1016/j.anucene.2017.03.022

[51] Meller, C.; Bremer, J.; Ankit, K.; Baur, S.; Bergfeldt, T.; Blum, P.; Canic, T.; Eiche, E.; Gaucher, E.; Hagenmeyer, V.; Heberling, F.;

Held, S.; Herfurth, S.; Isele, J.; Kling, T.; Kuhn, D.; Mayer, D.; Müller, B.; Nestler, B.; Neumann, T.; Nitschke, F.; Nothstein, A.; Nusiaputra, Y.; Orywall, P.; Peters, M.; Sahara, D.; Schäfer, T.; Schill, E.; Schilling, F.; Schröder, E.; Selzer, M.; Stoll, M.; Wiemer, H.-J.; Wolf, S.; Zimmermann, M.; Kohl, T., Integrated Research as Key to the Development of a Sustainable Geothermal Energy Technology. 2017. Energy technology, 5 (7), 965–1006. doi:10.1002/ente.201600579

[52] Miassoedov, A.; Albrecht, G.; Fluhrer, B.; Karbojian, A.; Konovalenko, A.; Kudinov, P.; Bechta, S., Results of the DEFOR-ROD1 test on the influence of BWR control rod guide tubes on melt jet fragmentation. 2017. Transactions of the American Nuclear Society, 117 (1), 913–916

[53] Miassoedov, A.; Fluhrer, B.; Journeau, C.; Bechta, S.; Hozer, Z.; Manara, D.; Bottomley, D.; Kiselova, M.; Keim, T.; Langrock, G.; Belloni, F., The safest project: severe accident facilities for European safety targets. 8th European Review Meeting on Severe Accident Research (ERMSAR), Warszawa, PL, May 16-18, 2017

[54] Miassoedov, A.; Fluhrer, B.; Journeau, C.; Bechta, S.; Hozer, Z.; Manara, D.; Bottomley, D.; Kiselova, M.; Keim, T.; Langrock, G.; Belloni, F.; Schyns, M., The safest project: towards Pan-European lab on Corium behavior in severe accidents. Main objectives and R&D priotities. 8th European Review Meeting on Severe Accident Research (ERMSAR), Warszawa, PL, May 16-18, 2017. Proceedings on USB-Stick, Paper 517

[55] Miassoedov, A.; Fluhrer, B.; Journeau, C.; Isaksson, P.; Bechta, S.; Hozer, Z.; Manara, D.; Bottomley, D.; Kiselova, M.; Keim, T.; Langrock, G.; Belloni, F., Experiments on BWR-specific severe accident phenomena in the European SAFEST project. Workshop on Advances in Understanding the Progression of Severe Accidnets in Boiling Water Reactors, Wien, A, July 17-21, 2017

[56] Miassoedov, A.; Gaus-Liu, X.; Journeau, C.; Liao, Y., ALISA project: access to large infrastructures for severe accidents in Europe and China. 17th International Topical Meeting on Nuclear Reactor Thermal Hydraulics, Xi'an, Shaanxi, China, September. 3-8, 2017

[57] Miassoedov, A.; Gaus-Liu, X.; Bechta, S.; Komlev, A.; Carenini, L.; Fichot, F., Heat transfer in homogeneous and stratified melt pools in the lower head of a reactor pressure vessel. Engineering Challenges for the Betterment of Society : 2nd thermal and Fluids Engineering Conference (TFEC), Las Vegas, NV, April 2-5, 2017

[58] Miassoedov, A.; Liao, Y.; Gaus-Liu, X.; Journeau, C.; Cassiaut-Louis, N.; Zhang, H.; Bai, W.; Zhang, Y. P.; Kuang, B.; Hu, P. ALISA project : Large-scale experimental facilities for severe accidents studies in Europe and China. International Congress on Advances in Nuclear Power Plants (ICAPP 2017), Fukui and Kyoto, Japan, 24th - 28th April 2017. Proceedings on CD-ROM, Art.Nr. 17217, Atomic Energy Society of Japan, Tokyo

[59] Mistrangelo, C.; Bühler, L., Instabilities in electrically driven rotating MHD layers. 2017. IOP conference series / Materials science and engineering, 228 (conference 1), 012004/1–10. doi:10.1088/1757-899X/228/1/012004

[60] Mistrangelo, C.; Bühler, L.; Brinkmann, H.-J., Combined experimental and numerical study of MHD flows in liquid metal blankets. 7th International Conference on Magneto-Science (ICMS 2017), Reims, F, October 23-27, 2017. Proceedings on USB-Stick, Paper No O4, 22–23, Unversity, Reims, F

[61] Mistrangelo, C.; Bühler, L.; Koehly, C.; Brinkmann, H.-J., Influence of modifications of HCLL blanket design on MHD pressure losses., 2017. Fusion engineering and design, 124, 948–952. doi:10.1016/j.fusengdes.2016.11.012

[62] Mistrangelo, C.; Bühler, L., Magnetohydrodynamic flows in liquid metal blankets for fusion reactors. 2017. Proceedings in applied mathematics and mechanics, 17 (1), 115–118. doi:10.1002/pamm.201710033

[63] Mistrangelo, C.; Bühler, L., MHD phenomena related to electromagnetic flow coupling. 2017. Magnetohydrodynamics, 53 (1), 141–148

[64] Mistrangelo, C.; Bühler, L., Magnetohydrodynamic flows in liquid metal blankets for fusion reactors. Jahrestagung der Gesellschaft für Angewandte Mathematik und Mechanik (GAMM 2017), Weimar, 6.-10.März 2017

[65] Münzberg, T.; Wiens, M.; Schultmann, F., A spatial-temporal vulnerability assessment to support the building of community resilience against power outage impacts. 2017. Technological Forecasting and Social Change, 121, 99–118. doi:10.1016/j.techfore.2016.11.027

[66] Orywall, P.; Drüppel, K.; Kuhn, D.; Kohl, T.; Zimmermann, M.; Eiche, E. Correction to: Flow-through experiments on the interaction of sandstone with Ba-rich fluids at geothermal conditions. 2017. Geothermal Energy, 5 (1), Art. Nr. 23. doi:10.1186/s40517-017-0083-y

[67] Orywall, P.; Drüppel, K.; Kuhn, D.; Kohl, T.; Zimmermann, M.; Eiche, E. Flow-through experiments on the interaction of sandstone with Ba-rich fluids at geothermal conditions. 2017. Geothermal Energy, 5 (1), Art. Nr.: 20. doi:10.1186/s40517-017-0079-7

[68] Otte, D.; Stoppel, L.; Wondrak, T.; Wetzel, T., Development of a measurement technique for bubble detection in liquid tin. 18th International UIE-Congress on Electrotechnologies for Material Processing, Hannover, June 6-9, 2017

[69] Ottenburger, S.; Airaksinen, M.; Pinto-Seppä, I.; Raskob, W., Enhancing urban resilience via a real-time decision support system for smart cities. 2017 International Conference on Engineering, Technology and Innovation (ICE/ITMC), Madeira Island, Portugal, 27–29 June 2017, 836–844, IEEE, Piscataway (NJ). doi:10.1109/ICE.2017.8279970

[70] Ottenburger, S.; Bai, S., Simulation based strategic decision making in humanitarian supply chain management. 2017 International Conference on Information and Communication Technologies for Disaster Management (ICT-DM), Münster, Germany, 11–13 December 2017, 1–7, IEEE, Piscataway (NJ). doi:10.1109/ICT-DM.2017.8275683

[71] Ottenburger, S.; Münzberg, T., An approach for analyzing the impacts of smart grid topologies on critical infrastructure resilience. Proceedings of the 14th International Conference on Information Systems for Crisis Response and Management (ISCRAM 2017). Agility is Coming. Albi, F, May 21-24, 2017. Ed.: T. Comes, 400–411, ISCRAM

[72] Ottenburger, S. S.; Münzberg, T.; Strittmatter, M., Smart Grid Topologies Paving the Way for an Urban Resilient Continuity Management. 2017. International Journal of Information Systems for Crisis Response and Management, 9 (4), 1–22. doi:10.4018/IJISCRAM.2017100101

[73] Pacio, J.; Daubner, M.; Fellmoser, F.; Hering, W.; Jäger, W.; Stieglitz, R.; Wetzel, T., Construction of experimental liquid metal facilities (experiences at KIT). Thermohydraulics and Chemistry of Liquid Metal Cooled Reactors, April 10-14, 2017. Ed.: F. Roelofs, Von Karman Inst. for Fluid Dynamics, Rhode Saint Genèse

[74] Pacio, J.; Litfin, K.; Wetzel, T.; Kennedy, G.; Van Tichelen, K., Thermal-hydraulic experiments supporting the MYRRHA fuel assembly. International Conference on Fast Reactors and Related Fuel Cycles: Next Generation Nuclear Systems for Sustainable Development (FR17), Yekaterinburg, Russia, June 26-29, 2017. Proceedings on USB-Stick, Paper IAEA-CN245–283

[75] Pacio, J.; Wetzel, T., Corrigendum to "Assessment of liquid metal technology status and research paths for their use as efficient heat transfer fluids in solar central receiver systems" [Solar Energy 93 (2013) 11–22]. 2017. Solar energy, 144, 827. doi:10.1016/j.solener.2017.02.040

[76] Palmiotti, G.; Salvatores, M.; Hursin, M.; Kodeli, I.; Gabrielli, F.; Hummel, A., New approaches to provide feedback from nuclear and covariance data adjustment for effective improvement of evaluated nuclear data files. 2017. The European physical journal / Web of Conferences, 146, 06003/1–11. doi:10.1051/epjconf/201714606003

[77] Pandazis, P.; Hollands, T.; Gaus-Liu, X.; Miassoedov, A., Experimental and numerical investigation of molten corium behavior in lower head under external subcooling and boiling conditions. 8th European Review Meeting on Severe Accident Research (ERMSAR), Warszawa, PL, May 16-18, 2017. Proceedings on USB-Stick, Paper 514

[78] Plevan, M., Entwicklung eines Verfahrens zur thermischen Zerlegung von Methan zu Wasserstoff und Kohlenstoff unter Nutzung flüssiger Metalle als Wärmeübertragungsmedium. Dissertation. 2017. Karlsruhe. doi:10.5445/IR/1000068714

[79] Raquet, M.; Class, A.; Edelbauer, W., Application of the Stochastic Field Method to two phase flow. 2017. Proceedings in applied mathematics and mechanics, 17 (1), 661–662. doi:10.1002/pamm.201710299

[80] Raquet, M.; Class, A.; Edelbauer, W., Application of the stochastic field method to two phase flow. Jahrestagung der Gesellschaft für Angewandte Mathematik und Mechanik (GAMM 2017), Weimar, 6.-10.März 2017

[81] Raskob, W.; Beresford, N., CONFIDENCE: Dealing with Uncertainties Relevant for Decision Making. 14th International Conference on Information Systems for Crisis Response and Management (ISCRAM 2017), Albi, France, May 21 - 24, 2017

[82] Raskob, W.; Möhrle, S., Nuclear Emergency Response and Big Data Technologies. Workshop on Big Data in Climate Action, Environment, Resource Efficiency and Raw Materials, Brussels, Belgium, November 6, 2017

[83] Raskob, W.; Münzberg, T., Schmutzige Bomben und ihre möglichen Folgen für die Bevölkerung. 2017. Crisis prevention, 1, 41–43

[84] Raskob, W.; Schichtel, T.; Landman, C. Das Entscheidungshilfesystem RODOS zur Online- und Realtime-Unterstützung. Behördenseminar "Notfallschutz außerhalb der Anlage", Gesellschaft für Anlagen- und Reak-torsicherheit (GRS), Köln, 15.-16. März 2017

[85] Reimann, J.; Vicente, J.; Brun, E.; Ferrero, C.; Gan, Y.; Rack, A., X-ray tomography investigations of mono-sized sphere packing structures in cylindrical containers. 2017. Powder technology, 318, 471–483. doi:10.1016/j.powtec.2017.05.033

[86] Ren, K.; Kotchourko, A.; Lelyakin, A.; Jordan, T., Numerical Reproduction of DDT in Small Scale Channels. 25th International Conference on Nuclear Engineering, Volume 8 : Computational Fluid Dynamics (CFD) and Coupled Codes - Nuclear Education, Public Acceptance and Related Issues, Shanghai, China, 2nd - 6th July 2017, Art.Nr. V008T09A039, ASME, New York (NY). doi:10.1115/ICONE25-67150

[87] Rineiski, A.; Flad, M.; Gabrielli, F.; Vezzoni, B., Fast reactor systems in the German P&T and related studies. International Conference on Fast Reactors and Related Fuel Cycles: Next Generation Nuclear Systems for Sustainable Development (FR17), Yekaterinburg, Russia, June 26-29, 2017.Proceedings on USB-Stick, Paper IAEA-CN245–487

[88] Rineiski, A.; Marchetti, M.; Andriolo, L.; Gabrielli, F., Recent neutronics developments

for reactor safety studies with SIMMER code at KIT. 2017. Journal of physics / Conference Series, 781 (1), 012036. doi:10.1088/1742-6596/781/1/012036

[89] Rineiski, A., Worldwide activities: Germany. 2017. Molten Salt Reactors and Thorium Energy. Ed.: T.J. Dolan, 651–654, Elsevier, Amsterdam

[90] Roelofs, F.; Shams, A.; Batta, A.; Moreau, V.; De Piazza, I.; Gerschenfeld, A.; Planquart, P.; Tarantino, M., The SESAME project: State of the art liquid metal thermal hydraulics and beyond. 2017. Atw, 62 (8/9), 531–537

[91] Sanchez-Espinoza, V. H.; Gomez-Garcia-Torano, I.; Rebollo, M. J.; Kretzschmar, F. ASTEC investigations of accident management measures for a station black-out (SBO) scenario in a generic German konvoi PWR. 8th European Review Meeting on Severe Accident Research (ERMSAR), Warszawa, PL, May 16-18, 2017. Proceedings on USB-Stick, Paper 609

[92] Schneider, T.; Gering, F.; Charron, S.; Zhelezniak, M.; Andronopoulos, S.; Heriard-Dubreuil, G.; Camps, J.; Raskob, W. Nuclear and Radiological Preparedness: The Achievements of the European Research Project PREPARE. 2017. Radiation protection dosimetry, 173 (1-3), 151–156. doi:10.1093/rpd/ncw318

[93] Schulenberg, T., Summary of SCWR Research in Europe, 8th International Symposium on Supercritical Water-Cooled Reactors (ISSCWR-8), Chengdu, China, March 13-15, 2017. Proceedings on USB-Stick, 1067

[94] Schulenberg, T., Review of severe accident management options for pressurized water reactors. 2nd Sino-German Symposium on Fundamentals of Advanced Nuclear Safety Technology (SG-FANS), Karlsruhe, September 12-15, 2017

[95] Schulenberg, T. (Hrsg.)., Annual Report 2016 of the Institute for Nuclear and Energy Technologies (KIT Scientific Reports ; 7742). 2017. KIT Scientific Publishing, Karlsruhe. doi:10.5445/KSP/1000072764

[96] Schulenberg, T.; Li, H., Transient heat transfer in an out-of-pile SCWR fuel assembly test at near-critical pressure. 8th International Symposium on Supercritical Water-Cooled Reactors (ISSCWR-8), Chengdu, China, March 13-15, 2017. Proceedings on USB-Stick, 390–405

[97] Schulenberg, T.; Xiao, J., Nuclear energy research at Karlsruhe Institute of Technology. 2nd KIT Innovation Day, Suzhou, China, October 18, 2017

[98] Schwenk-Ferrero, A.; Andrianov, A., Nuclear Waste Management Decision-Making Support with MCDA. 2017. Science and technology of nuclear installations, 2017, 1–20. doi:10.1155/2017/9029406

[99] Schwenk-Ferrero, A., Decision making support applying multi-criteria decision aid methods to wicked waste management problems. Research on Radioactive Waste Management : Ethics - Society - Technology, Final ENTRIA Conference, Braunschweig, September 26-30, 2017

[100] Schwenk-Ferrero, A.; Andrianov, A., Comparison and Screening of Nuclear Fuel Cycle Options in View of Sustainable Performance and Waste Management. 2017. Sustainability, 9 (9), Art.Nr. 1623. doi:10.3390/su9091623

[101] Staudt, C.; Raskob. W., Modellierung des Radionuklidtransfers in einem europäischen Entscheidungshilfeprogramm. 2. Workshop - Helmholtz Cross Program Activity Querschnittsthema Strahlenforschung "Transportprozesse in Mensch und Umwelt", GSI Helmholtzzentrum für Schwerionenforschung, Darmstadt, 24.-25.10.2017

[102] Stoppel, L.; Fehling, T.; Geißler, T.; Baake, E.; Wetzel, T., Carbon dioxide free pro-

duction of hydrogen. 2017. IOP conference series / Materials science and engineering, 228 (1), Art. Nr.: 012016. doi:10.1088/1757-899X/228/1/012016

[103] Stoppel, L.; Geißler, T.; Wetzel, T., Process concept and lab-scale test of CO_2-free methane cracking via liquid metal technology. Electrotechnologies for Material Processing: 18th International UIE-Congress, Hannover, June 6-9, 2017. Ed.: E. Baake, 411–415, Vulkan-Verl., Essen

[104] Talamo, A.; Gohar, Y.; Gabrielli, F.; Rineiski, A.; Pyeon, C. H., Advances in the computation of the Sjöstrand, Rossi, and Feynman distributions. 2017. Progress in nuclear energy, 101, 299–311. doi:10.1016/j.pnucene.2017.01.006

[105] Van Dorsselaere, J.-P.; Brechignac, F.; De Rosa, F.; Herranz, L. E.; Kljenak, I.; Miassoedov, A.; Paci, S.; Piluso, P., Trends in severe accident research in Europe: SARNET network from Euratom to NUGENIA. 2017. EPJ Nuclear Sciences & Technologies, 3, 28/1–10. doi:10.1051/epjn/2017021

[106] van Dorsselaere, J. P.; Brechignac, F.; De Rosa, F.; Herranz, L. E.; Kljenka, I.; Miassoedov, A.; Paci, C.; Piluso, P. Trends in severe accident research in Europe: SARNET network from EURATOM to NUGENIA. Topical Meeting on Safety in Reactor Operations (TopSafe 2017), Wien, A, February 12-16, 2017.Proceedings publ. online

[107] Vezzoni, B.; Marchetti, M.; Andriolo, L.; Gabrielli, F.; Chen, X. N.; Matzerath-Boccaccini, C.; Rineiski, A.; Maschek, W.; Morita, K.; Arima, T., SIMMER analyses of the EBR-II shutdown heat removal tests. International Conference on Fast Reactors and Related Fuel Cycles: Next Generation Nuclear Systems for Sustainable Development (FR17), Yekaterinburg, Russia, June 26-29, 2017. Proceedings on USB-Stick, Paper IAEA-CN245–072

[108] Vezzoni, B.; van Rooijen, W. F. G.; Fei, T.; Zhang, Y.; Marchetti, M.; Balestra, P.; Parisi, C., IAEA neutronics benchmark for EBR-II SHRT-45R. International Conference on Fast Reactors and Related Fuel Cycles: Next Generation Nuclear Systems for Sustainable Development (FR17), Yekaterinburg, Russia, June 26-29, 2017. Proceedings on USB-Stick, Paper IAEA-CN245–070

[109] Wang, S.; Massone, M.; Rineiski, A.; Merle-Lucotte, E.; Laureau, A.; Gerardin, D.; Heuer, D.; Allibert, M., A passive decay heat removal system for emergency draining tanks of molten salt reactors. 17th International Topical Meeting on Nuclear Reactor Thermal Hydraulics Xi'an, Shaanxi, China, Sept. 3-8, 2017. Proceedings on USB-Stick, Paper No.19773

[110] Wei, H.; Erkan, N.; Okamoto, K.; Gaus-Liu, X.; Miassoedov, A., Improvement and evaluation of debris coolability analysis module in severe accident analysis code SAMPSON using LIVE experiment. 2017. Annals of nuclear energy, 100, 65–72. doi:10.1016/j.anucene.2016.10.011

[111] Wiemer, H.-J.; Kuhn, D.; Barredo, S. Use of binary cycles for geothermal heat sources from diverse geological settings. Geologia, Presente y Futuro : XX Congreso Geologico Argentino, Tucuman, RA, August 7-11, 2017 Hrsg.: R. Manoni, 84–85, published online

[112] Xiao, J.; Breitung, W.; Kuznetsov, M.; Zhang, H.; Travis, J. R.; Redlinger, R.; Jordan, T., GASFLOW-MPI: A new 3-D parallel all-speed CFD code for turbulent dispersion and combustion simulations. Part I: Models, verification and validation. 2017. International journal of hydrogen energy, 42 (12), 8346–8368. doi:10.1016/j.ijhydene.2017.01.215

[113] Xiao, J.; Breitung, W.; Kuznetsov, M.; Zhang, H.; Travis, J. R.; Redlinger, R.; Jordan, T., GASFLOW-MPI: A new 3-D parallel all-speed CFD code for turbulent dispersion and combustion simulations Part II: First analysis of

the hydrogen explosion in Fukushima Daiichi Unit 1. 2017. International journal of hydrogen energy, 42 (12), 8369–8381. doi:10.1016/j.ijhydene.2017.01.219

[114] Xiao, J., Development and Validation of the Parallel All-Speed CFD Code GASFLOW-MPI for Detonation of Proemixed H_2-Air Mixture in a Hemispherical Balloon. 25th International Conference on Nuclear Engineering (ICONE 25), Shanghai, China, May 14-18, 2017. Proceedings on CD-ROM, Paper ICONE25–66066, ASME, New York (NY)

[115] Xiao, J.; Breitung, W.; Kuznetsov, M.; Travis, J.; Redlinger, R., Development and Validation of the Parallel All-Speed CFD Code GASFLOW-MPI for Detonation of Premixed H_2-Air Mixture in a Hemispherical Balloon. 25th International Conference on Nuclear Engineering, Volume 8: Computational Fluid Dynamics (CFD) and Coupled Codes: Nuclear Education, Public Acceptance and Related Issues, Shanghai, China, 2nd - 6th July 2017, Art.Nr. V008T09A003, ASME, New York (NY). doi:10.1115/ICONE25-66066

[116] Xiao, J.; Breitung, W.; Kuznetsov, M.; Zhang, H., Numerical investigations of turbulent slow deflagration of premixed H_2-air-H_2O mixture in THAI test HD-22 using CFD code GASFLOW-MPI. 17th International Topical Meeting on Nuclear Reactor Thermal Hydraulics, Xi'an, Shaanxi, China, Sept. 3-8, 2017, Paper No. 20323

[117] Xiao, J.; Travis, J.; Bottoni, M. Status of dynamic water film model development in 3-D CFD code GASFLOW-MPI for analysis of passive containment cooling system. 2017. Annals of nuclear energy, 108, 99–112. doi:10.1016/j.anucene.2017.04.007

[118] Yu, F.; Class, A. G.; Xiao, J.; Jordan, T. Coarse grid CFD methodology: flux corrections for individual mesh cells and application to rod bundles [in press]. 17th International Topical Meeting on Nuclear Reactor Thermal Hydraulics, Xi'an, Shaanxi, China, Sept. 3-8, 2017, Paper No. 21717

[119] Yu, J.; Hou, B.; Lelyakin, A.; Xu, Z.; Jordan, T. L., Gas detonation cell width prediction model based on support vector regression. 2017. Nuclear engineering and technology, 49 (7), 1423–1430. doi:10.1016/j.net.2017.06.014

[120] Zhang, H.; Guo, J.; Lu, J.; Niu, J.; Li, F. A comparison of coupling algorithms for N/TH transient problems in HTR. International Conference on Mathematics and Computational Methods Applied to Nuclear Science and Engineering (M&C 2017), Jeju, Korea, April 16-20, 2017. Proceedings on USB-Stick, P354S09–01ZhangH

[121] Zhang, H.; Li, Y.; Xiao, J.; Jordan, T. L. Large eddy simulation of turbulent flow using the parallel computational fluid dynamics code GASFLOW-MPI. 2017. Nuclear engineering and technology, 49 (6), 1310–1317. doi:10.1016/j.net.2017.08.003

[122] Zhang, H.; Xiao, J.; Jordan, T. Large Eddy simulation of turbulent flows using the CFD code GASFLOW-MPI. International Conference on Mathematics and Computational Methods Applied to Nuclear Science and Engineering (M&C 2017), Jeju, Korea, April 16-20, 2017. Proceedings on USB-Stick, P326S07–01ZhangH

[123] Zhang, H.; Xiao, J.; Li, Y.; Jordan, T. A numerical study of the release and dispersion of a light gas using 3-d CFD code GASFLOW-MPI. 7th International Conference on Hydrogen Safety (ICHS-2017), Hamburg, September 11-13, 2017

[124] Zhang, X.; Huang, M., Ensemble-based release estimation for accidental river pollution with known source position. 2017. Journal of hazardous materials, 333, 99–108. doi:10.1016/j.jhazmat.2017.03.028

[125] Zhang, X.; Raskob, W.; Landman, C.; Trybushnyi, D.; Haller, C.; Yuan, H., Automatic plume episode identification and cloud shine reconstruction method for ambient gamma dose rates during nuclear accidents. 2017. Journal of environmental radioactivity, 178-

179, 36–47. doi:10.1016/j.jenvrad.2017.07.014

[126] Zhang, X.; Raskob, W.; Landman, C.; Trybushnyi, D.; Li, Y., Sequential multi-nuclide emission rate estimation method based on gamma dose rate measurement for nuclear emergency management. 2017. Journal of hazardous materials, 325, 288–300. doi:10.1016/j.jhazmat.2016.10.072

Publications 2018

[1] Albrecht, G.; Miassoedov, A.; Meyer, L.; Meignen, R., Ex-vessel fuel coolant interaction experiment in the DISCO facility in the SAFEST project. 12th International Topical Meeting on Nuclear Reactor Thermal-Hydraulics, Operation and Safety (Nuthos-12), Qingdao, China, October, 14-18, 2018. Procedings on CD-ROM, Paper 1157

[2] Arlt, T., Analyse des Stabilitätsverhaltens magnetohydrodynamischer Kanalströmungen. Dissertation. 2018. Karlsruhe. doi:10.5445/IR/1000085856

[3] Bai, S.; Staudt, C.; Raskob, W. OPERRA-HARMONE, knowledge database and FDMT regionalization. 4th NERIS Workshop (2018), Dublin, Irland, 25.–27. April 2018

[4] Bouyer, V.; Denois, A.; Journeau, C.; Molina, D.; Piluso, P.; Haquet, J. F.; Foit, J.; Fluhrer, B.; Miassoedov, A., Molten core concrete interaction test in VULCANO facility preventing initial interfacial crusts. 12th International Topical Meeting on Nuclear Reactor Thermal-Hydraulics, Operation and Safety (Nuthos-12), Qingdao, China, October, 14-18, 2018. Proceedings on USB-Stick, Paper 919

[5] Bühler, L.; Brinkmann, H. J.; Koehly, C. Experimental confirmation of additional magnetohydrodynamic pressure drop at junctions between electrically insulating flow channel inserts. MHD Days and GdRI Dynamo Meeting, Dresden, November 26-28, 2018

[6] Bühler, L.; Aiello, G.; Bendotti, S.; Koehly, C.; Mistrangelo, C.; Galabert, J., Development of combined temperature – Electric potential sensors. 2018. Fusion engineering and design, 136, 7–11. doi:10.1016/j.fusengdes.2017.12.004

[7] Bühler, L.; Brinkmann, H. J.; Köhly, C. Experimental study of liquid metal magnetohydrodynamic flows near gaps between flow channel inserts. 30th International Symposium on Fusion Technology (SOFT 2018), Giardini-Naxos, Italien, 16.–21. September 2018

[8] Bühler, L.; Mistrangelo, C., Pressure drop and velocity changes in MHD pipe flows due to a local interruption of the insulation. 2018. Fusion engineering and design, 127, 185–191. doi:10.1016/j.fusengdes.2018.01.010

[9] Bühler, L.; Mistrangelo, C.; Brinkmann, H.-J.; Koehly, C., Pressure distribution in MHD flows in an experimental test-section for a HCLL blanket. 2018. Fusion engineering and design, 127, 168–172. doi:10.1016/j.fusengdes.2018.01.007

[10] Büttner, F.; Heiler, W.; Gabriel, S.; Kuhn, D., Experimentelle Untersuchung zur Blasenentstehung durch einen vertikalen, in ein Flüssigbad eintauchenden Freistrahl. Fachtagung 'Experimentelle Strömungsmechanik', Rostock, 4.-6.September 2018

[11] Dietrich, B.; Laube, T.; Marocco, L.; Wetzel, T., Heat transfer of liquid metal flow in a tube heated on half of the circumference - concept of a test loop. 16th International Heat Transfer Conference (IHTC-16), Beijing, China, August 10-15, 2018. Proceedings on USB-Stick, Paper IHTC16–24226

[12] Fichot, F.; Carénini, L.; Sangiorgi, M.; Hermsmeyer, S.; Miassoedov, A.; Bechta, S.;

Zdarek, J.; Guenadou, D., Some considerations to improve the methodology to assess In-Vessel Retention strategy for high-power reactors. 2018. Annals of nuclear energy, 119, 36–45. doi:10.1016/j.anucene.2018.03.040

[13] Foit, J. J., MOCKA experiments with and without rebars. Concrete decomposition temperature relevant to MCCI process. Crusts and concrete ablation. 2nd Management Board Meeting, OECD/NEA Project on Thermodynamic Characterisation of Fuel Debris and Fission Products Based on Scenario Analysis of Severe Accident Progression at the Fukushima Daiichi Nuclear Power Station (TCOFF), Paris, F, January 25-17, 2018

[14] Foit, J. J.; Cron, T.; Fluhrer, B.; Miassoedov, A.; Isaksson, P.; Bechta, S. MOCKA-SSM experiments on basaltic concrete with and without rebars. 12th International Topical Meeting on Nuclear Reactor Thermal-Hydraulics, Operation and Safety (Nuthos-12), Qingdao, China, October, 14-18, 2018. Proceedings on USB-Stick, Paper 917

[15] Foit, J. J.; Fluhrer, B.; Cron, T.; Miassoedov, A., MOCKA-SSM experiments with and without rebars. 5th NUGENIA-SARNET TA2.2 Review Meeting on Ex-Vessel Corium Interactions and Coolability, Puerto de la Cruz, E, April 18-20, 2018

[16] Freitag, M.; Kljenak, I.; Jankowski, T.; Risken, T.; Kostka, P.; Götz, L.; Klauck, M.; Schwarz, S.; Bleyer, A.; Bentaib, A.; Povilaitis, M.; Siccama, N. B.; Royl, P.; Mansour, A.; Janda, T., Benchmark exercise TH-27 on natural convection with steam injection and condensation inside the extended THAI facility. 2018. Annals of nuclear energy, 116, 90–104. doi:10.1016/j.anucene.2018.02.027

[17] Gabrielli, F.; Sánchez-Espinoza, V. H. Uncertainty and Sensitivity Analysis by Means of ASTEC/URANIE Platform of the QUENCH-08 Experiment. Proceedings of the 24th International Quench Workshop, Karlsruhe, November 13-15, 2018, 354–375

[18] Gabriel, S.; Schulenberg, T.; Albrecht, G.; Heiler, W.; Miassoedov, A.; Kaiser, F.; Wetzel, T., Optical void measurement method for stratified wavy two phase flows. 2018. Experimental thermal and fluid science, 97, 341–350. doi:10.1016/j.expthermflusci.2018.05.001

[19] Gaus-Liu, X.; Cron, T.; Fluhrer, B.; Miassoedov, A.; Fichot, F., Experimental study on the melt pool transient behaviour in LWR lower head under changing boundary cooling conditions: The outcomes of LIVE-SAFEST experiment. 12th International Topical Meeting on Nuclear Reactor Thermal-Hydraulics, Operation and Safety (Nuthos-12), Qingdao, China, October, 14-18, 2018. Proceedings on USB-Stick, Paper 1163

[20] Gaus-Liu, X.; Cron, T.; Fluhrer, B.; Miassoedov, A.; Zhang, L.; Xu, J.; Zhang, H. Experimental study on the melt pool transient behaviour in LWR lower head under changing boundary cooling conditions: The outcomes of LIVE-ALISA experiment. 12th International Topical Meeting on Nuclear Reactor Thermal-Hydraulics, Operation and Safety (Nuthos-12), Qingdao, China, October, 14-18, 2018. Proceedings on USB-Stick, Paper 1083

[21] Gaus-Liu, X.; Cron, T.; Fluhrer, B.; Miassoedov, A.; Wenz, T., Preliminary results of LIVE2D 2-layer experiment. 3rd IVMR (In-Vessel Melt Retention Strategy for High Power Nuclear Reactors) Annual Meeting, Praha, CZ, May 28-31, 2018

[22] Gaus-Liu, X.; Cron, T.; Fluhrer, B.; Miassoedov, A.; Wenz, T., Updates of LIVE3D and LIVE2D experiments. Nuclear Days and NUGENIA Annual Forum, Prahe, CZ, April 10-12, 2018

[23] Guo, L.; Flad, M.; Rineiski, A., Debris behavior and FCI effects n the core catcer of an ESFR reactor under accident conditions. 12th International Topical Meeting on Nuclear Reactor Thermal-Hydraulics, Operation and Safety (Nuthos-12), Qingdao, China, October, 14-18, 2018. Proceedings on USB-Stick, Paper 908

[24] Guo, L.; Rineiski, A., Numerical Investigation of Corium Coolability in Core Catcher: Sensitivity to Modeling Parameters. 26th International Conference on Nuclear Engineering, London, England, July 22–26, 2018. Vol.: 4, ICONE26–81841, ASME, New York (NY). doi:10.1115/ICONE2681841

[25] Haas, C.; Kaiser, F.; Schulenberg, T.; Wetzel, T., Critical heat flux for flow boiling of water on micro-structured Zircaloy tube surfaces. 2018. International journal of heat and mass transfer, 120, 793–806. doi:10.1016/j.ijheatmasstransfer.2017.12.077

[26] Han, Z.; Yabing, L.; Jianjun, X.; Jordan, T., Preliminary validation of the detached eddy simulation model in CFD code GASFLOW-MPI. 26th International Conference on Nuclear Engineering, ICONE 2018; London; United Kingdom; 22 July 2018 through 26 July 2018, V008T09A044, ASME, New York City (NY). doi:10.1115/ICONE26-82402

[27] Höhne, T.; Gabriel, S., Simulation of a counter-current horizontal gas-liquid flow experiment at the WENKA channel using a droplet entrainment model. 2018. Annals of nuclear energy, 121, 414–425. doi:10.1016/j.anucene.2018.07.047

[28] Holbein, B.; Schulenberg, T. Investigation on refrigerant transport by capillary effect with fleeces In an evaporator for a high temperature cooling machine [Étude sur le transport de frigorigène par effet de capillarité avec des molletons dans un évaporateur pour une machine de refroidissement à haute température]. 2018. International journal of refrigeration, 93, 18–28. doi:10.1016/j.ijrefrig.2018.05.036

[29] Jordan, T., Safe use of LH2 in non-industrial seeting. New work item proposal. 27th Plenary Meeting of the ISO TC 197 (2018), Vancouver, Kanada, 6.–7. Dezember 2018

[30] Jordan, T., Overview of development on hydrogen safety analysis code at KIT. Open Seminar on Advanced Hydrogen Safety for Nuclear Plants of the METI/JAEA (2018), Tokio, Japan, 11. Dezember 2018

[31] Jordan, T., Sektorenkopplung mit Wasserstoff als 'Treibstoff'. 2.Öffentliches Kolloquium der Forschungspartnerschaft KA Energie, Karlsruhe, 20.November 2018

[32] Jordan, T., Hydrogen safety. University of Adelaide, AUS, October 29, 2018

[33] Jordan, T., Update on KIT work on H_2 safety issues. IEA HydrogenTask 37 Meeting, Paris-Saclay, F, October 19, 2018

[34] Jordan, T., PRESLHY Project Presentation. IEA HydrogenTask 37 Meeting, Paris-Saclay, F, October 19, 2018

[35] Jordan, T., Hydrogen safety. 5th International Energy Symposium, Karlsruhe, September 11-12, 2018

[36] Jordan, T.; Franzen, J.; Graf, F.; Griffith, A.; Hansen, O.; Haberzettl, A.; Hawksworth, S.; Weinberger, B.; Wilde, P., Update of Hydrogen Safety Research Priorities in the Application Domain. Research Priorities Workshop, Buxton, GB, September 18-20, 2018

[37] Jordan, T.; Moretto, P.; Dolci, F.; Keller, J., Report on the IA HySafe / EU-JRC / US-DOE H_2 Safety Research Priorities Workshop RPW2016. 22nd World Hydrogen Energy Conference, Rio de Janeiro, BR, June 17-22, 2018

[38] Jorden, T., Technical issues on hydrogen safety analyses and analytical models for large-scale hydrogen analyses of diffusion, combustion and explosion. Meeting on Advanced Nuclear Hydrogen Safety (2018), Tokio, Japan, 12. Dezember 2018

[39] Jordan, T., HyTunnel and KIT contributions. Tunnel Workshop of the IPHE Regulations, Codes, Standards and Safety Working Group, Buxton, GB, September 18, 2018

[40] Kaiser, F.; Gabriel, S.; Albrecht, G.; Wetzel, T., Investigation of the critical heat flux

in a rod bundle configuration under low pressure conditions. 2018. Ideen und Innovationen für die Energie von morgen : Wissenschaftliche Beiträge des KIT zu den Jahrestagungen 2014, 2016 und 2017 des KIT-Zentrums Energie. Ed.: W. Breh, 37–41, KIT Scientific Publishing, Karlsruhe

[41] Kashinsky, O.; Kulikov, D.; Kurdyumov, A.; Lezhnin, S.; Lobanov, P.; Pribaturin, N.; Svetonosov, A.; Pacio, J.; Stoppel, L.; Wetzel, T., Influence of Spacer Elements on Flow Distribution and Heat Transfer in Experimental Models of Fuel Assemblies. 26th International Conference on Nuclear Engineering, London, England, July 22–26, 2018. Vol.: 6B, ICONE26–82163, ASME, New York (NY). doi:10.1115/ICONE26-82163

[42] Kennedy, G.; Van Tichelen, K.; Pacio, J.; Di Piazza, I.; Uitslag-Doolaard, H., Thermal-hydraulic experimental testing of the MYRRHA wire-wrapped fuel assembly. Advances in Thermal Hydraulics (ATH 2018) embedded in ANS 2018 Winter Meeting, Orlando, FL, November 11-15, 2018, 505–519, ANS, La-Grange Park (IL)

[43] Khabensky, V. B.; Granovsky, V. S.; Almjashev, V. I.; Vitol, S. A.; Krushinov, E. V.; Kotova, S. J.; Sulatsky, A. A.; Gusarov, V. V.; Bechta, S. V.; Barrachin, M.; Bottomley, D.; Fischer, M.; Hellmann, S.; Piluso, P.; Miassoedov, A.; Tromm, W., Effect of temperature gradient on chemical element partitioning in corium pool during in-vessel retention., 2018. Nuclear engineering and design, 327, 82–91. doi:10.1016/j.nucengdes.2017.11.030

[44] Kim, H. Y.; Bechta, S.; Foit, J.; Hong, S. W., In-Vessel and Ex-Vessel Corium Stabilization in Light Water Reactor., Science and technology of nuclear installations, 2018, 1–3. doi:10.1155/2018/3918150

[45] Kivva, S.; Zheleznyak, M.; Boyko, O.; Ievdin, I.; Pylypenko, O.; Mikhalsky, O.; Raskob, W.; Sorokin, M., Updated module of radionuclide hydrological dispersion of the decision support system RODOS., EGU General Assembly (2018), Wien, Österreich, 8.–13. April 2018

[46] Klüber, V.; Bühler, L.; Mistrangelo, C., Numerical simulations of 3D magnetohydrodynamic flows in dual-coolant lead lithium blankets., 30th International Symposium on Fusion Technology (SOFT 2018), Giardini-Naxos, Italien, 16.–21. September 2018

[47] Klüber, V.; Bühler, L.; Mistrangelo, C. Pressure-driven 3D liquid metal MHD flow in a dual-coolant lead lithium blanket for a fusion reactor. MHD Days and GdRI Dynamo Meeting, Dresden, November 26-28, 2018

[48] Köhly, C.; Bühler, L.; Mistrangelo, C. Design of a test section to analyze magneto-convection effects in WCLL blankets. 23rd Topical Meeting on the Technology of Fusion Energy (TOFE 2018), Orlando, FL, November 11-15, 2018. Proceedings on USB-Stick, Paper No. 732–18

[49] Köhly, C.; Neuberger, H.; Bühler, L. Fabrication of the thin-walled fusion blanket components like flow channel inserts by selective laser manufacturing. 30th International Symposium on Fusion Technology (SOFT 2018), Giardini-Naxos, Italien, 16.–21. September 2018

[50] Krasnov, D.; Bandaru, V.; Bühler, L.; Boeck, T., Instabilities and turbulence in magnetohydrodynamic duct flows. NIC Symposium 2018, Jülich, February 22-23, 2018

[51] Krasnov, D.; Boeck, T.; Bühler, L., Turbulent and transitional sidewall jets in magnetohydrodynamic channels with a homogeneous magnetic field. 2018. Proceedings in applied mathematics and mechanics, 17 (1), 111–114. doi:10.1002/pamm.201710032

[52] Kuznetsov, M., Hydrogen flame propagation regimes and flame instabilities in a thin layer compartment. High-Speed Combustion and Explosion Seminar (2018), Yokohama, Japan, 12. November 2018

[53] Kuznetsov, M., Hydrogen flame propagation regimes and flame instabilities in a thin layer compartment. Gaseous Detonation Seminar (2018), Tokio, Japan, 5. November 2018

[54] Kuznetsov, M., Large scale experiments on hydrogen combustion with respect to safety of nuclear power plants. Gaseous Detonation Seminar (2018), Tokio, Japan, 5. November 2018

[55] Kuznetsov, M., Safety aspects of high pressure alkaline electrolyzers for hydrogen production. KIT-Toshiba Co-Operation Seminar (2018), Fuchu, Japan, 13. November 2018

[56] Kuznetsov, M., Large scale experiments on hydrogen combustion with respect to safety of nuclear power plants. KIT-Toshiba Co-Operation Seminar (2018), Fuchu, Japan, 13. November 2018

[57] Kuznetsov, M.; Chaumeix, N.; Yanez, J.; Tengah, S., Flame folding and wrinkling factor for 2D and 3D hydrogen-air flames. 2018. Joint Meeting of the German and Italian Sections of the Combustion Institute, 41st Meeting of the Italian Section of the Combustion Institute, Sorrent, Italien, 23.–26. Mai 2018 Proceedings. Ed.: H. Bockhorn, 593–598, ASICI, Napolı

[58] Kuznetsov, M.; Grune, J., Hydrogen flame propagation regimes in a thin layer compartment. 15th International Conference on Flow Dynamics (ICFD 2018), Sendai, J, November 7-11, 2018. Proceedings on USB-Stick, 364–365, Institute of Fluid Science, Tohoku University

[59] Kuznetsov, M.; Yanez Escancano, J.; Grune, J., A shock wave - combustion front ensemble propagating with the Chapman-Jouguet velocity being of subsonic nature. 37th International Symposium on Combustion, Dublin, IRL, July 29 - August 3, 2018

[60] Kuznetsov, M.; Yanez, J.; Bykov, V. Shockless flame acceleration and DDT in open channels driven by hydraulic resistance. Joint Meeting of the German and Italian Sections of the Combustion Institute, 41st Meeting of the Italian Section of the Combustion Institute. Sorrent, Italien, 23.–26. Mai 2018. Proceedings. Ed.: H. Bockhorn, 532–537, ASICI, Napoli

[61] Li, Y.; Zhang, H.; Xiao, J.; Travis, J. R.; Jordan, T., Numerical investigation of natural convection inside the containment with recovering passive containment cooling system using GASFLOW-MPI. 2018. Annals of nuclear energy, 114, 1–10. doi:10.1016/j.anucene.2017.11.047

[62] Li, Y.; Xiao, J.; Jordan, T., Numerical study of turbulent natural convection in an air filled square cavity with GASFLOW-MPI code. 12th International Topical Meeting on Nuclear Reactor Thermal-Hydraulics, Operation and Safety (Nuthos-12), Qingdao, China, October, 14-18, 2018. Proceedings on CD-ROM, Paper 1097

[63] Li, Y.; Xiao, J.; Zhang, H., Validation of film evaporation model in GASFLOW-MPI. 2018 26th International Conference on Nuclear Engineering, ICONE 2018; London; United Kingdom; 22 July 2018 through 26 July 2018, 6 p., ASME, New York City (NY). doi:10.1115/ICONE2681249

[64] Madokoro, H., Thermo-structural analysis of a reactor pressure vessel lower head during core-melt severe accidents. Dissertation. 2018. Karlsruhe. doi:10.5445/IR/1000087838

[65] Madokoro, H.; Miassoedov, A.; Schulenberg, T., A thermal structural analysis tool for RPV lower head behavior during severe accidents with core melt. 2018. Bulletin of the JSME, 4, 18–00038. doi:10.1299/mel.18-00038

[66] Madokoro, H.; Miassoedov, A.; Schulenberg, T., Assessment of a lower head molten pool analysis module using LIVE experiment. 2018. Nuclear engineering and design, 330, 51–58. doi:10.1016/j.nucengdes.2018.01.036

[67] Makarov, D.; Hooker, P.; Kuznetsov, M.; Molkov, V., Deflagrations of localised homogeneous and inhomogeneous hydrogen-air mixtures in enclosures. 2018. International journal of hydrogen energy, 43 (20), 9848–9869. doi:10.1016/j.ijhydene.2018.03.159

[68] Makarov, D.; Shentsov, V.; Kuznetsov, M.; Molkov, V., Pressure peaking phenomenon: Model validation against unignited release and jet fire experiments. 2018. International journal of hydrogen energy, 43 (19), 9454–9469. doi:10.1016/j.ijhydene.2018.03.162

[69] Marchetti, M.; Rineiski, A., Minor actinide balance reduction in ESFR-SMART. 2018. Journal of physics / Conference Series, 1133 (1), 012027. doi:10.1088/1742-6596/1133/1/012027

[70] Maschek, W.; Li, R.; Matzerath Boccaccini, C.; Gabrielli, F.; Morita, K., Investigation on upper bounds of recriticality energetics of hypothetical core disruptive accidents in sodium cooled fast reactors. 2018. Nuclear engineering and design, 326, 392–402. doi:10.1016/j.nucengdes.2017.11.002

[71] Massone, M. V. E., Cross-Sections for Transient Analyses: Development of a Genetic Algorithm for the Energy Meshing. Dissertation. 2018. Karlsruhe. doi:10.5445/IR/1000083680

[72] Massone, M.; Wang, S.; Rineiski, A.; Servell, P., Analytical modeling of the emergency draining tank for a molten salt reactor. 4th International Conference on Physics and Technology of Reactors and Applications (PHYTRA4), Marrakech, MA, September 17-19, 2018. Proceedings on CD-ROM, 475–486

[73] Mercan, A. K.; Sánchez-Espinoza, V. H.; Gabrielli, F., Validation of ASTEC2.1 using QUENCH-12 for VVER-reactors. Proceedings of the 24th International Quench Workshop, Karlsruhe, November 13-15, 2018, 376–406

[74] Miassoedov, A.; Albrecht, G.; Cron, T.; Gaus-Liu, X.; Fluhrer, B.; Wenz, T., Experimental Studies of Core Melt Behavior During Severe Accidents at the Karlsruhe Institute of Technology. 2018. Transactions of the American Nuclear Society, 119, 904–907

[75] Miassoedov, A.; Fluhrer, B.; Journeau, C.; Bechta, S.; Hózerd, Z.; Manara, D.; Kiselova, M.; Keim, T.; Langrock, G.; Belloni, F.; Bottomley, D.; Kurata, M.; Okamoto, K. SAFEST project on corium behavior in severe accidents. 12th International Topical Meeting on Nuclear Reactor Thermal-Hydraulics, Operation and Safety (Nuthos-12), Qingdao, China, October, 14-18, 2018. Proceedings on USB-Stick, Paper 912

[76] Miassoedov, A.; Fluhrer, B.; Journeau, C.; Bechta, S.; Hózerd, Z.; Manara, D.; Kiselova, M.; Keim, T.; Langrock, G.; Belloni, F.; Bottomley, D.; Kurata, M.; Okamoto, K. DEFOR test on the influence of BWR control rod guide tubes on melt jet fragmentation. 12th International Topical Meeting on Nuclear Reactor Thermal-Hydraulics, Operation and Safety (Nuthos-12), Qingdao, China, October, 14-18, 2018. Proceedings on USB-Stick, Paper 913

[77] Miassoedov, A.; Gaus-Liu, X.; Liao, Y.; Chen, P., ALISA project: Access to large infrastructures for severe accidents in Europe and China. 12th International Topical Meeting on Nuclear Reactor Thermal-Hydraulics, Operation and Safety (Nuthos-12), Qingdao, China, October, 14-18, 2018. Proceedings on USB-Stick, Paper 911

[78] Mistrangelo, C.; Bühler, L.; Brinkmann, H.-J., Experimental investigation of MHD pressure losses in a mock-up of a liquid metal blanket. 2018. Nuclear fusion, 58 (3), Art.Nr. 036012. doi:10.1088/1741-4326/aaa133

[79] Moehrle, S.; Bai, S.; Mueller, T.; Munz, E.; Trybushnyi, D.; Raskob, W., Web-based decision support system for emergency man-

agement. System architecture and enhancement possibilities. 4th NERIS Workshop (2018), Dublin, Irland, 25.–27. April 2018

[80] Münzberg, T.; Ottenburger, S. S., Schutz kritischer Infrastrukturen: Kritikalität als Entscheidungsmaß zur Abwehr von Gefahre am Beispiel Stromausfall. 2018. Was heißt Kritikalität? : Zu einem Schlüsselbegriff der Debatte um kritische Infrastrukturen. Hrsg.: J.I. Engels, 179–214, transcript Verl., Bielefeld

[81] Neuberger, H.; Rey, J.; Hernandez, F.; Ruck, S.; Arbeiter, F.; Rieth, M.; Koehly, C.; Stratil, L.; Niewoehner, R.; Felde, A. Evaluation of conservative and innovative manufacturing routes for gas cooled Test Blanket Module and Breeder Blanket First Walls. 30th International Symposium on Fusion Technology (SOFT 2018), Giardini-Naxos, Italien, 16.–21. September 2018

[82] Niedermeier, K.; Marocco, L.; Flesch, J.; Mohan, G.; Coventry, J.; Wetzel, T., Performance of molten sodium vs. molten salts in a packed bed thermal energy storage. 2018. Applied thermal engineering, 141, 368–377. doi:10.1016/j.applthermaleng.2018.05.080

[83] Ottenburger, S. S.; Cakmak, H. K.; Jakob, W.; Blattmann, A.; Deines, E.; Trybushnyi, D.; Raskob, W.; Hagenmeyer, V., Smart Grid resilience - security of supply 2.0. 7. Jahrestagung des KIT-Zentrums Energie (2018), Karlsruhe, Deutschland, 26. Juni 2018

[84] Ottenburger, S. S.; Daniell, J., Direct and indirect vulnerability analysis framework for decentralized power systems. International Conference on 'Natural Hazards and Risks in a Changing World', Potsdam, October 4-5, 2018

[85] Pacio, J., Thermal-hydraulic experiments with liquid metals—Introduction. 2018. Thermal Hydraulics Aspects of Liquid Metal Cooled Nuclear Reactors. Ed.: F. Roelofs, 45–82, Elsevier Science & Technology, San Diego. doi:10.1016/B978-0-08-101980-1.00003-X

[86] Pacio, J.; Daubner, M.; Fellmoser, F.; Litfin, K.; Wetzel, T., Heat transfer experiment in a partially (internally) blocked 19-rod bundle with wire spacers cooled by LBE. 2018. Nuclear engineering and design, 330, 225–240. doi:10.1016/j.nucengdes.2018.01.034

[87] Pacio, J.; Daubner, M.; Wetzel, T., Local blockages in a rod bundle with wire spacers: Heat transfer in LBE for the safety assessment of MYRRHA. Advances in Thermal Hydraulics (ATH 2018) embedded in ANS 2018 Winter Meeting, Orlando, FL, November 11-15, 2018, 520–530, ANS, LaGrange Park (IL)

[88] Pacio, J.; Daubner, M.; Wetzel, T.; Di Piazza, I.; Tarantino, M.; Martelli, D.; Angelucci, M., Experimental Nusselt Number in Rod Bundles Cooled by Heavy-Liquid Metals. 26th International Conference on Nuclear Engineering, London, England, July 22–26, 2018. Vol.: 6B, ICONE26–82213, ASME, New York (NY). doi:10.1115/ICONE26-82213

[89] Pacio, J.; Niedermeier, K.; Wetzel, T., Liquid metal technology for high temperature heat transport and storage. Power-Heat-Power : International Workshop on Carnot Batteries, Stuttgart, October 9-10, 2018

[90] Pandazis, P.; Hollands, T.; Gaus-Liu, X.; Miassoedov, A., Experimental and numerical investigation of molten corium behavior in lower head under external subcooling and boiling conditions. 2018. Annals of nuclear energy, 120, 888–895. doi:10.1016/j.anucene.2018.06.020

[91] Payot, F.; Journeau, C.; Suteau, C.; Serre, F.; Gradeck, M.; RImbert, N.; Lecoanet, A.; Miassoedov, A., A new experimental R&D program associated with the corium jet impingement on the ASTRID core catcher sacrificial material. International Congress on Advances in Nuclear Power Plants (ICAPP 2018), Charlotte, NC, April 8-11, 2018

[92] Raquet, M.; Class, A.; Edelbauer, W. Application of the stochastic field method to cavitating flows in an injection nozzle.

10th International Symposium on Cavitation (CAV2018), Baltimore, MD, May 14-16, 2018. Proceedings

[93] Raquet, M.; Class, A. G.; Edelbauer, W. Application of the stochastic field method to multi-fluid flow. 89. Jahrestagung der Gesellschaft für angewandte Mathematik und Mechanik (GAMM 2018), München, Deutschland, 19.–23. März 2018

[94] Raskob, W., Emergency preparedness: advancement and still open gaps in Europe. 5th European IRPA Congress (2018), Den Haag, Niederlande, 4.–8. Juni 2018

[95] Raskob, W.; Almahayni, T.; Beresford, N. A., Radioecology in CONFIDENCE: Dealing with uncertainties relevant for decision making. 2018. Journal of environmental radioactivity, 192, 399–404. doi:10.1016/j.jenvrad.2018.07.017

[96] Rineiski, A.; Meriot, C.; Marchetti, M.; Krepel, J., Core safety measures in ESFR-SMART. PHYSOR 2018 : Reactor Physics Paving the Way Towards More Efficient Systems, Cancun, MEX, April 22-26, 2018

[97] Rineiski, A.; Sinitsa, V., C4P-train neutronics tool for supporting safety studies of innovative fast reactors. 4th International Conference on Physics and Technology of Reactors and Applications (PHYTRA4), Marrakech, MA, September 17-19, 2018. Proceedings on CD-ROM, 679–688

[98] Schöttl, P.; Bern, G.; van Rooyen, D. W.; Flesch, J.; Fluri, T.; Nitz, P., Efficient modeling of variable solar flux distribution on Solar Tower receivers by interpolation of few discrete representations. 2018. Solar energy, 160, 43–55. doi:10.1016/j.solener.2017.11.028

[99] Schröder, E.; Thomauske, K.; Kuhn, D.; Jodocy, M.; Kraml, M.; Thorwart, K.; Wolfgramm, M., Experimentelle Bestimmung der physikalischen Thermalwassereigenschaften. Der Geothermiekongress DGK 2018, Essen, 27.-29.November 2018, Paper F6.4

[100] Schröder, E.; Thomauske, K.; Kuhn, D.; Thorwart, K.; Nowak, K.; Wolfgramm, M. BMWi-Verbundprojekt PEther "Physikalische Eigenschaften von Thermalwasser unter in-situ-Bedingungen", Teilvorhaben 0325761A: "Experimentelle Bestimmung der physikalischen Eigenschaften von Thermalwasser in situ und im Labor" : Abschlussbericht : Laufzeit des Vorhabens: 01.03.2015 bis 31.12.2017. 2018. Karlsruher Institut für Technologie. doi:10.2314/GBV:1029410496

[101] Schulenberg, T., Hydrogen combustion in combined cycle power plants. 5th International Energy Symposium, Karlsruhe, September 11-12, 2018

[102] Schulenberg, T., Special Issue : Selected Papers from the 8th International Symposium on Supercritical Water-Cooled Reactors. 2018. Journal of Nuclear Engineering and Radiation Science, 4 (1), Art.Nr 010301. doi:10.1115/1.4037556

[103] Schulenberg, T.; Li, H., Transient heat transfer in an out-of-pile SCWR fuel assembly test at near-critical pressure. 2018. Journal of Nuclear Engineering and Radiation Science, 4 (1), Art.Nr. 011005. doi:10.1115/1.4038061

[104] Schulenberg, T.; Starflinger, J.; Leung, L., Design of SCWR. Internatioal Training Course Generation IV : Nuclear ReactorSystems for the Future, Saclay, F, June 18-22, 2017

[105] Schumm, T.; Frohnapfel, B.; Marocco, L., Investigation of a turbulent convective buoyant flow of sodium over a backward- facing step. 2018. Heat and mass transfer, 54 (8), 2533–2543. doi:10.1007/s00231-017-2102-8

[106] Utili, M.; Boccaccini, L.; Buehler, L.; Cismondi, F.; DelNevo, A.; Eboli, M.; DiFonzo, F.; Hernandez, T.; Wulf, S.-E.; Kordac, M.; Martinelli, D.; Masdelesvalls, E.; Melichar, T.; Mistrangelo, C.; Reungoat, M.; Tincani, A.

Development of Pb-16Li technologies for DEMO reactor. 30th International Symposium on Fusion Technology (SOFT 2018), Giardini-Naxos, Italien, 16.–21. September 2018

[107] Wang, S.; Muscher, H.; ASTEC-Team at KIT., Preliminary analysis of a BWR SBO SA-sequence with ASTEC2.1rev3. 8th ASTEC Users' Club Meeting, Aix-en-Provence, F, October 8-11, 2018

[108] Weisenburger, A.; Hering, W.; Onea, A.; Pacio, J.; Schroer, C.; Wetzel, T.; Stieglitz, R.; Müller, G., Liquid Metals – a Promising Heat Transfer Media for CSP – Overview on KIT Research. SolarPACES (2018), Casablanca, Marokko, 2.–5. Oktober 2018

[109] Wiemer, H.-J., Niedertemperaturkreislauf Karlsruhe. Projekt MoNiKa. 6.Praxisforum Geothermie Bayern, München, 17.Oktober 2018. Hrsg.: J. Schneider, 28, Enerchange GmbH, München

[110] Wiens, M.; Raskob, W.; Diehlmann, F.; Wandler, S.; Schultmann, F., Exposure and vulnerability of the energy system to internal and external effects. 2018. Energy as a Sociotechnical Problem. Ed. by Christian Büscher, Jens Schippl, Patrick Sumpf., 96–122, Routledge, London

[111] Xiao, J.; Kuznetsov, M.; Travis, J. R. Experimental and numerical investigations of hydrogen jet fire in a vented compartment. 2018. International journal of hydrogen energy, 43 (21), 10167–10184. doi:10.1016/j.ijhydene.2018.03.178

[112] Yu, F.; Zhang, H.; Li, Y.; Xiao, J.; Class, A.; Jordan, T., Voxelization-based high-efficiency mesh generation method for parallel CFD code GASFLOW-MPI. 2018. Annals of nuclear energy, 117, 277–289. doi:10.1016/j.anucene.2018.03.045

[113] Zhang, H.; Guo, J.; Li, F.; Xu, Y.; Downar, T. J., Efficient simultaneous solution of multi-physics multi-scale nonlinear coupled system in HTR reactor based on nonlinear elimination method. 2018. Annals of nuclear energy, 114, 301–310. doi:10.1016/j.anucene.2017.12.014

[114] Zhang, H.; Guo, J.; Lu, J.; Li, F.; Xu, Y.; Downar, T. J., The Improvement of Coupling Method in TINTE by Fully Implicit Scheme. 2018. Rheumatologia, 190 (2), 156–175. doi:10.1080/00295639.2018.1426299

[115] Zhang, H.; Li, Y.; Xiao, J.; Jordan, T. Detached Eddy Simulation of hydrogen turbulent dispersion in nuclear containment compartment using GASFLOW-MPI. 2018. International journal of hydrogen energy, 43 (29), 13659–13675. doi:10.1016/j.ijhydene.2018.05.077

[116] Zhang, H.; Li, Y.; Xiao, J.; Jordan, T., Uncertainty analysis of condensation heat transfer benchmark using CFD code GASFLOW-MPI. 2018. Nuclear engineering and design, 340, 308–317. doi:10.1016/j.nucengdes.2018.10.007

[117] Zhang, H.; Li, Y.; Xiao, J.; Jordan, T. Large eddy simulations of the all-speed turbulent jet flow using 3-D CFD code GASFLOW-MPI. 2018. Nuclear engineering and design, 328, 134–144. doi:10.1016/j.nucengdes.2017.12.032

[118] Zhang, L.; Xu, J.; Wie, D.; Zhang, H.; Gaus-Liu, X.; Miassoedov, A.; Cron, T.; Fluhrer, B., Experimental investigations of IVR natural circulation characteristics and CHF: 2DIVR ALISA test. 12th International Topical Meeting on Nuclear Reactor Thermal-Hydraulics, Operation and Safety (Nuthos-12), Qingdao, China, October, 14-18, 2018. Proceedings on USB-Stick, Paper 929